Digital Electronics Projects for Beginners

Digital Electronics Projects for Beginners

Owen Bishop

PC Publishing

PC Publishing
4 Brook Street
Tonbridge
Kent TN9 2PJ

First published 1990

© Owen Bishop

ISBN 1 870775 08 2

All rights reserved. No part of this publication may be reproduced or transmitted in any form, including photocopying and recording, without the written permission of the copyright holder, application for which should be addressed to the Publishers. Such written permission must also be obtained before any part of this publication is stored in an information retrieval system of any nature.

The book is sold subject to the Standard Conditions of Sale of Net Books and may not be re-sold in the UK below the net price given by the Publishers in their current price list.

British Library Cataloguing in Publication Data
Bishop, O.N. (Owen Neville),
   Digital electronics projects for beginners
   1. Digital electronics equipment
   I. Title
   621. 3815
   ISBN 1-870775-08-2

Phototypesetting by Scribe Design
Printed in Great Britain by BPCC Wheatons Ltd, Exeter

Contents

Introduction – What is digital electronics?   1

1   Logic probes   9

2   Movie show   16

3   Progressive timer   24

4   Intruder detector   33

5   Capacitance meter   41

6   Combination door sentry   50

7   Digital die   59

8   A Christmas decoration   66

9   Weekly reminder   72

10   Remote-control switch   78

11   Metronome   89

12   Anemometer   96

Appendix: Notes for beginners   109

Index   118

Introduction – What is digital electronics?

Logic is concerned with reasoning, with working things out. In this book we build electronic circuits that operate according to logical rules.

In logic, statements are either true or they are untrue. Therefore, if we are to build circuits capable of performing logic, we must find ways of representing 'true' and 'untrue' in such circuits. There are several ways of doing this, as these examples show:

Represented by	If true	If untrue
a switch	switch closed	switch open
a lamp	lamp on	lamp off
a voltage	high	low (or zero)
a motor	running	not running
a buzzer	sounding	not sounding

All of these operate on an 'all-or-nothing' basis. The switch is either closed or open; it can not be half-open. There can be no 'half-truths'. The lamp is either on or off. Although a lamp *can* be half on, it is not allowed to be so in a logic circuit, since this would be a 'half-truth'. Half-truths are meaningless. Similarly, the motor is either running or not running.

The two-state nature of truth and untruth can also be represented numerically. Usually, 'true' is represented by the digit '1' and 'untrue' is represented by '0'. This is when logic becomes *digital logic*.

Introduction – What is digital electronics?

Logic gates

The basic circuit units from which most digital logic circuits are built are the logic *gates*. A logic gate may be quite a complicated circuit, consisting of several transistors, diodes and resistors. Luckily, you do not have to build the logic gates yourself. They come ready-built as complete units in the form of integrated circuits, as will be explained later. Better still, there is no need to know *how* a gate works. All you need to know is *what it does*. An example of a logic gate is the AND gate, which performs the logical AND operation. One version of this gate, as shown by its symbol in Fig 0.1, has two input terminals. It also has an output terminal. It also has two terminals for the positive and negative power supply though, since *all* gates need a power supply, the symbol does not show this.

Figure 0.1 Symbol for the logic gate AND

To use a gate, we connect its power terminals to the power supply and then apply a high voltage or a low voltage to each input terminal. As a result, the output terminal shows a high or low voltage. The state of the output voltage (high or low) depends upon the state of the voltages at the inputs, as decided by the logical rules of the gate.

Let us see what happens in the case of the 2-input AND gate. The gate can have a high voltage or a low voltage at each terminal. Thus there are four possible combinations of input, as shown in this table:

Inputs		Output
A	B	Z
0	0	0
0	1	0
1	0	0
1	1	1

In the table '1' represents a 'high' input voltage and corresponds to 'true'. Conversely, '0' represents a 'low' input voltage and

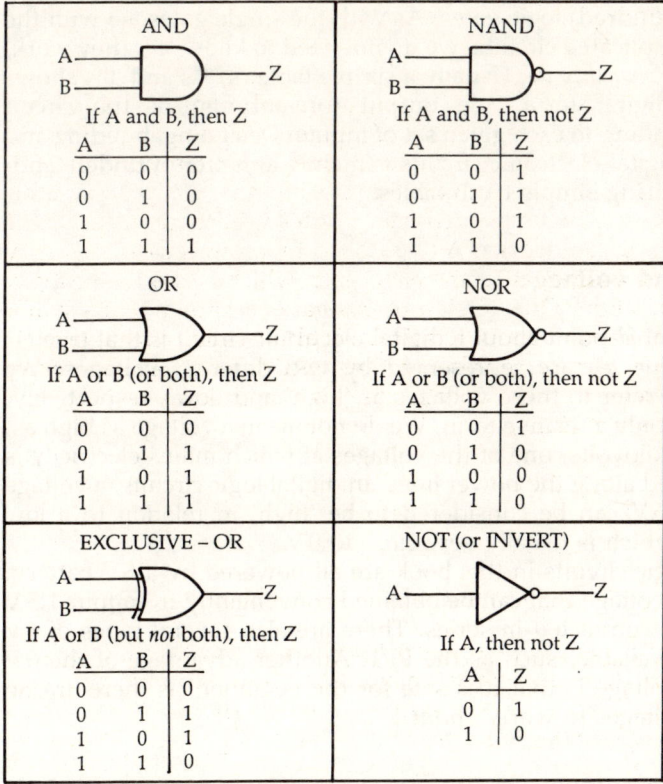

Figure 0.2 Symbols, logical statements and truth tables of the main types of logic gates

corresponds to 'untrue'. The table shows that the output of an AND gate is 'high' (true) only when input A is high (true) AND input B is high (true). It performs the logic of the sentence 'If A AND B, then Z'.

A table such as that above, which shows the truth relationships of the inputs and the output, is known as a *truth table*. This truth table tells us all we need to know about the behaviour of the AND gate. There are several other types of logic gate, each with their own truth tables. Fig 0.2 shows the symbols and explains the logic of these gates. There are also logic gates that have three or more inputs; we shall explain their action as the need arises. We shall also use several logic circuits built up of two or more (perhaps

Introduction – What is digital electronics?

several hundred) logic gates. As with the single gates, so with the more complicated circuits, we do not need to know *how* they work, but only *what they do*. Usually a simple table of '1's and '0's shows us, in digital form, the output (or outputs) of the circuit corresponding to each given set of inputs. Designing, building and testing digital electronic circuits is mainly a matter of understanding and using simple truth tables.

Truth and voltage

The essential point about a digital electronic circuit is that true (1) and untrue (0) are represented by two different voltages. We normally refer to these voltages as 'high' and 'low', respectively. 'High' is only a relative term. We do not mean a voltage as high as, say, 400 kilovolts, one of the voltages at which mains electricity is distributed along the power lines. In digital logic circuits, a voltage such as 6 V can be considered to be 'high' in relation to a low voltage which is usually very close to 0 V.

The logic circuits in this book are all powered by a 6 V battery. This is a voltage that can be obtained conveniently from four 1.5 V dry cells, connected in series. There are also several types of 6 V battery available, such as the PP1. Another advantage of the 6 V supply voltage is that it is safe for the beginner, as there are no mains voltages to worry about.

Integrated circuits

There are two series of logic integrated circuits that can be operated from a 6 V supply. These are the 74HC series and the 4000 series. The various integrated circuits in these series are designated by type numbers such as 74HC00, 74HC13, 74HC93, and so on for the 74HC series. The 74HC08, for example, consists of four 2-input AND gates. The corresponding device in the 4000 series is the 4081. All integrated circuits in the 4000 series have numbers in the 'four thousand and something' range, such as 4002, 4011, 4046, 4511 and so on.

Both the 74HC series and the 4000 series employ a semiconductor technology known as CMOS. The advantage of this technology is that it requires very little current, making batteries last longer. In both series, a low output from a logic gate is a voltage between 0 V and 0.05 V. A high output is a voltage between 5.95 V and 6 V.

Thus there is a big difference between a low and high output. For all practical purposes, we can take low as being 0 V and high as being 6 V. The specifications of the series leaves a margin for error for input voltages, so that operation of complicated logic circuits is entirely reliable. Any voltage between 0 V and about 1.3 V counts as a low input. Any voltage between 4.3 V and 6 V counts as a high input.

Although we take 6 V as the standard supply voltage for this book, the ics can operate on other voltages. The 74HC series operates on any voltage in the range 3 V to 6 V, while the 4000 range operates on 3V to 18 V. With a 6 V supply, we can mix ics of both series in the same circuit, as their output and input voltages are compatible.

All the integrated circuits we use in this book are manufactured on a tiny chip of silicon, perhaps only 1 millimetre square. To make it possible to handle this minute circuit, and to make the electrical connections to it, the chip is enclosed in a plastic package with a row of terminal pins down either side. This type of package is known as a *double-in-line* (d.i.l.) package. Fig 0.3 shows the system of numbering the pins. The end at which pin 1 is located is identified by a notch, a dot or occasionally by a bar printed on the top of the package. Fig 0.3 also shows, as an example, the connections to the gates of the 74HC08 ic, which contains four

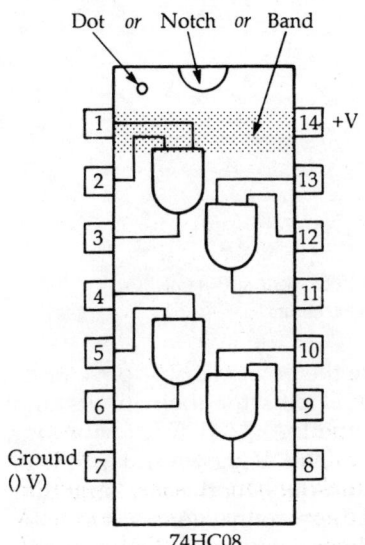

Figure 0.3 Integrated circuit with four AND gates, in 74HC series

Introduction – What is digital electronics?

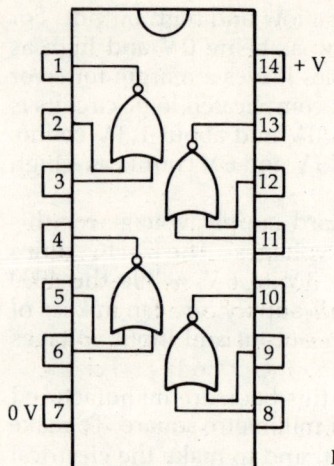

74HC02

Figure 0.4 Pin-out of NOR gates in 74HC series

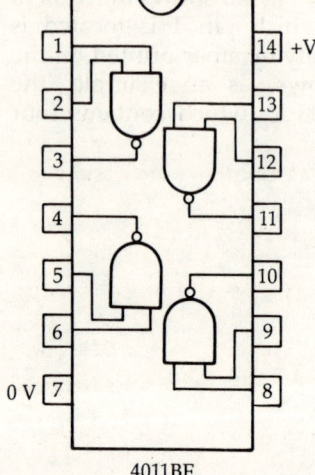

4011BE

Figure 0.5 Pin-out of NAND (or other gates) in 4000 series

2-input AND gates. Pins 7 and 14 are the power supply pins for all four gates. The gates have individual pins for their inputs and outputs. The connections are the same for all 74HC ics with four 2-input gates, except for the 74HC02. This is shown in Fig 0.4. In Fig 0.5 we show the connections (or pin-outs) for the corresponding ics of the 4000 series. Note that all these connections are seen as from *above* the ic when inserted in its socket on the circuit-board.

Handling ics

One of the problems with CMOS ics is that they are liable to damage by high static voltages. When you walk across a nylon carpet, your body becomes charged to a potential of several hundred or even thousands of volts. If you touch the pins of a CMOS ic the current discharged through the ic may be sufficient to damage the transistors inside. For this reason, CMOS ics need to be handled with care. In industry, a workstation may be equipped with special conductive mats, and devices to earth the operator and equipment. At home we can not take such extreme precautions but a few simple rules should be followed:

(1) Wear clothing made from natural fibres, such as cotton and wool, to minimise the charge built up when your body and clothes rub together.

(2) Until you need to use them, keep all CMOS ics with their pins embedded in conductive foam or in the metallised cartons in which they are purchased.

(3) If possible, work on a bare metal surface. For example, use the upturned lid of a biscuit 'tin', and wire the lid to earth (e.g. connect it to a cold-water pipe).

(4) Earth your body and tools immediately before handling the ics or the circuit board. The simplest way is to have on the work-bench a table-lamp or other piece of equipment (or the 'tin' lid mentioned above) which has exposed metal that is connected to mains earth. Briefly touch the earthed metal with your fingers or with metal tools from time to time. Discharge from a pointed tool such as a screwdriver or drill is particularly rapid and damaging, so this precaution is an important one.

There are two other rules which you should bear in mind if you are testing partly-completed circuits:

(1) Voltages must never be applied to input terminals unless the power terminals are already connected and the power supply is on.

(2) All input terminals must be connected to something. The ics can not be guaranteed to work properly if there are any unconnected inputs.

Logic and numbers

We have used the digits '0' and '1' to represent 'true' and 'untrue', but they can also be used to represent numerical values. In everyday life we express numbers by using the digits '0' to '9'. We

Introduction – What is digital electronics?

use these ten digits to express numbers in our familiar *decimal* system. Another number system is the *binary* system. This expresses numbers using only the two digits '0' and '1'.

This table shows the binary numbers that correspond to the first 16 numbers of the decimal system:

Decimal	Binary	Decimal	Binary
0	0	8	1000
1	1	9	1001
2	10	10	1010
3	11	11	1011
4	100	12	1100
5	101	13	1101
6	110	14	1110
7	111	15	1111

Thus, the '0's and '1's of numbers expressed in the binary system can be represented in electronic circuits by low and high voltage levels. Since arithmetic involves purely logical operations, it is possible to design logic circuits that will handle numbers. We can build circuits to count and circuits to perform mathematical operations, such as addition and subtraction. Such circuits are the basis of electronic computing devices from the humblest pocket calculator to the most sophisticated and powerful mainframe. We shall not be building such complex calculating devices in this book, but several of the projects involve using logic circuits to count and to perform very simple calculations, such as division.

Resistors

In the projects, use either 0.25 W carbon resistors (5% tolerance) or 0.6 W metal film resistors (1% tolerance) unless other types are specified.

1 Logic probes

This project is really two small projects in one. If you are a complete beginner, start with these. They illustrate some important points about digital electronics and they are useful for testing and trouble-shooting the projects you will build later.

The two kinds of probe are a logic level detector and a pulse detector (Fig 1.1). The level detector is used to check outputs from logic gates, to see if they are high or low. This is used when a gate has a steady or slowly changing output level. Sometimes we need to check a level that appears only for a few microseconds. This is where the pulse detector in needed.

How they work

First of all we consider the logic level detector. A small current flows through D1, R1 and D2 in Fig 1.1(a). It is a property of a diode that there is a fairly fixed voltage drop across it when a current flows through it. In the case of the diodes in these circuits, which are silicon diodes, this *forward voltage drop* is 0.7 V. Therefore, point B in Fig 1.1a is at 0.7 V, while point A is 0.7 V below the supply voltage. If the supply is +6 V, point A is at +5.3 V. However, if we run the circuit on +12 V, point A is at +11.3V. Whatever the supply voltage, point A is close to supply voltage and is effectively a logic high. Point B is close to 0 V and is logic low.

IC1 is an operational amplifier, wired as a comparator. It compares the voltage at point A with the voltage at the probe tip. If the voltage at the probe tip is *higher* than the voltage at A (i.e. higher than +5.3 V) the output of the amplifier is high. A current

Logic probes

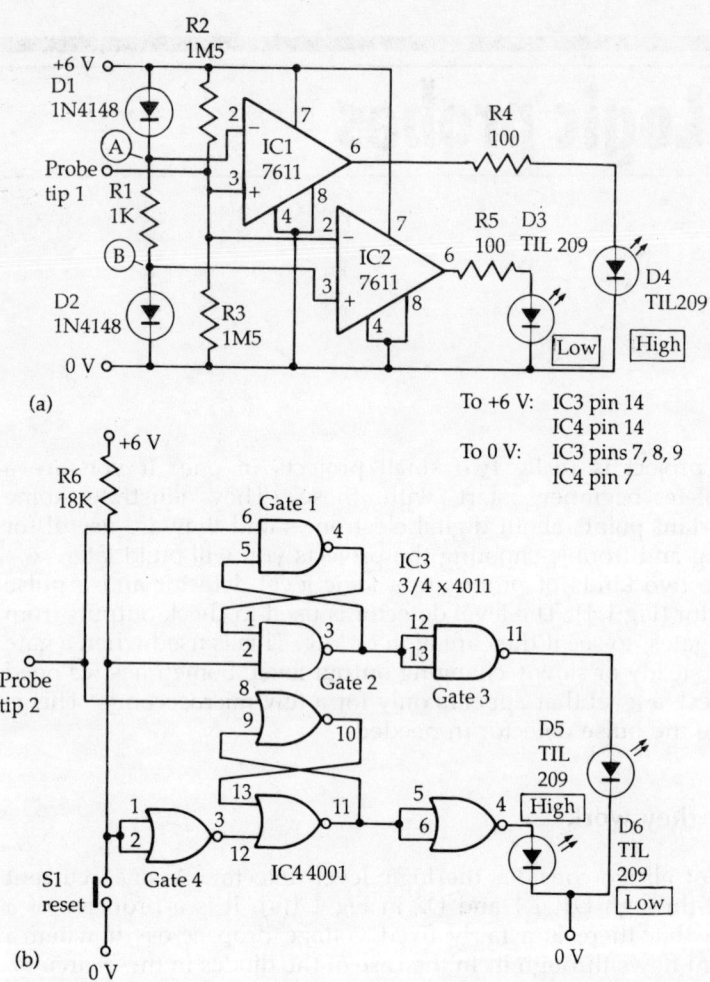

Figure 1.1 Logic probes; (a) level detector, (b) pulse detector

flows through R4 and the light-emitting diode D4. The diode comes on, indicating that there is logic high at the probe tip. This is what happens when the probe tip is touched against a terminal of the tested circuit at logic high. If the probe tip is not in contact with the tested circuit, the resistors R2 and R3, hold the probe tip at a voltage half of the supply (3 V) and the LED does not light. IC2 is another comparator but wired the other way round. The LED D3

10

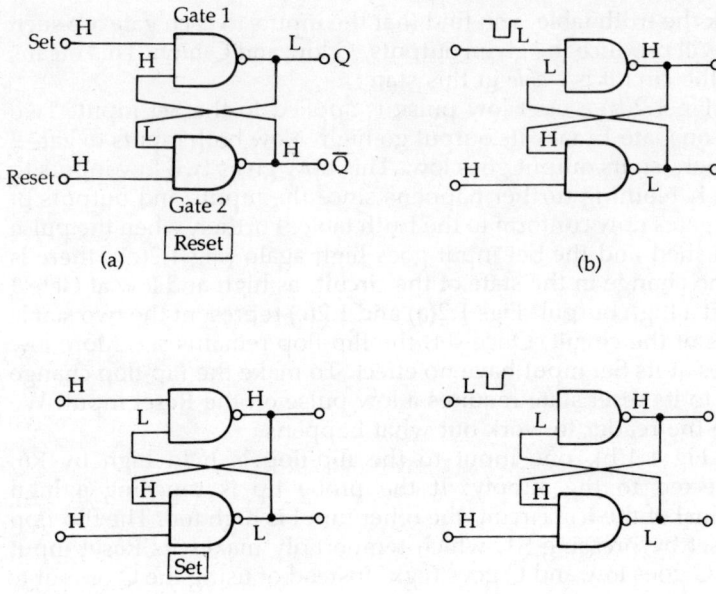

Figure 1.2 How a flip-flop works

lights when the voltage of the probe is *lower* than the voltage at B (i.e. less than 0.7 V). Thus we have two LEDS, one lights to indicate 'high', the other lights to indicate 'low', and neither lights when the voltage is intermediate or if the probe tip is not in contact with a live point.

The level detector is not a digital logic circuit, strictly speaking, but the pulse detector certainly is. It makes use of an arrangement of two logic gates that we shall often see in other circuits. This is the *bistable* circuit, so called because it is stable in either one of two states. It is often called a *flip-flop*. Fig 1.1(b) includes a flip-flop made from two NAND gates, labelled Gate 1 and Gate 2. Fig 1.2 shows how it works. Refer to the truth table for NAND on p. 3, to follow its operation. In Fig 1.2(a) the state of the inputs and outputs is indicated by H (=high) and L (=low). The two inputs to the flip-flop are called Set and Reset. They are both being held high, by means not shown in Fig 1.2. The outputs of the flip-flop are called Q and $\bar{Q}$. The 'bar' over the $\bar{Q}$ indicates INVERT, since, as we shall see, $\bar{Q}$ is always the INVERT or opposite of Q. If you

11

Logic probes

check the truth table, you find that the inputs to each gate are such that will produce the given outputs, Q low and $\bar{Q}$ high. This means that the circuit is *stable* in this state.

In Fig 1.2(b) a brief low pulse is applied to the Set input. Two lows on Gate 1 make its output go high. Now both inputs to gate 2 are high, so its output goes low. This now gives two low inputs to Gate 1. Nothing further happens since the inputs and outputs of both gates now conform to the truth table. Further, when the pulse is finished and the Set input goes high again (Fig 1.2(c)), there is still no change in the state of the circuit, as high and low at Gate 1 give it a high output. Figs 1.2(a) and 1.2(c) represent the two stable states of the circuit. Once set, the flip-flop remains set. More low pulses at its Set input have no effect. To make the flip-flop change back to its reset state requires a low pulse on the Reset input. We leave the reader to work out what happens.

In Fig 1.1(b), one input to the flip-flop is held high by R6, connected to the supply. If the probe tip is touching a high terminal of the test circuit, the other input is high too. The flip-flop is reset by pressing S1, which temporarily makes its Reset input low. Q goes low and $\bar{Q}$ goes high. Instead of using the Q output to drive the LED directly, which would make the flip-flop less sensitive, we take the output $\bar{Q}$, and invert it using Gate 3. The probe tip is normally connected to a high level output, and S1 is pressed to reset the flip-flop. The LED goes out. A brief low pulse detected by the probe tip sets the flip-flop and the LED comes on.

The other flip-flop in Fig 1.1(b) is based on two NOR gates. Its mode of operation is the reverse of that of the NAND gate flip-flop. It requires its Set and Reset inputs to be normally *low*, and is made to change state by a *high* pulse. Gate 4 is wired as an INVERT to supply a low resetting input of the flip-flop. A high pulse at the probe tip causes the LED to light.

Construction

You can build each probe separately, or combine them into one unit. If you build them separately, each uses its own probe and you have the advantage of being able to look at logic levels in one part of a circuit and to watch for pulses in another part. If you combine them, they can share a single probe. There are no problems in construction, except to remember that the operational amplifiers (IC1 and IC2) are CMOS devices and therefore need to be handled with care (p. 7). Also, it is advisable to employ a heat shunt when soldering the diodes (p. 112).

Construction

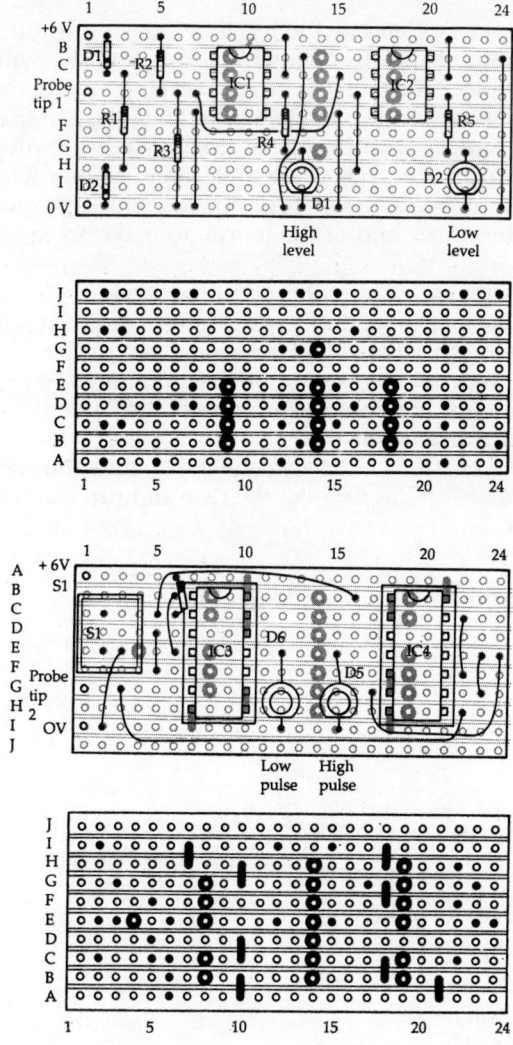

Figure 1.3 Logic probes; stripboard layouts (a) (b)

The circuits take their power from the circuit being tested. Solder lengths of flexible wire to the power supply terminals with crocodile clips at the free ends. The probe or probes can take various forms. You may decide to make the probe from an old ball-pen. Find one that had a *metal* tip (the cheaper ones have plastic that *looks like* metal). Remove the ink tube (care!). Strip the insulation from a length of thin flexible wire and thread the wire through the barrel. Replace the metal tip in the barrel, jamming the bare end of the wire between the barrel and the metal tip, so that the bare end makes contact with the metal tip. Replace the plug that is usually at the upper end of the barrel, jamming the wire between the plug and the barrel to secure it. It is convenient to glue the circuit board to the upper end of the barrel, to make an easily manipulated unit.

Instead of, or perhaps as well as, the pen-type probe, you can use a small crocodile clip, a miniature clip-on probe or a spring-loaded test prod. These are ideal for connecting to terminal pins of ics and leave your hands free for other operations.

The stripboard layouts are intended for use with the board exposed. If you prefer to enclose the board in a plastic case, fit terminal pins instead of the LEDs and the reset push-button, mount these components on the panel of the case and run leads to them from the board.

Components required for the level detector

Resistors (see p. 8)
R1   1k
R2,R3   1M5 (2 off)
R4,R5   100 (2 off)

Semiconductors
D1,D2   1N4148 silicon signal diode (2 off)
D3,D4   TIL209 or similar light emitting diodes (2 off)

Integrated circuits
IC1, IC2 7611 CMOS operational amplifier (2 off)

Miscellaneous
Stripboard 63 mm x 25 mm (Vero 14354)
8-pin d.i.l. sockets (2 off)
1 mm terminal pins (3 off)
miniature crocodile clips, insulated, (1 red, 1 black)
materials or clips for probe

Components required for the pulse detector

Resistors (see p. 8)
R6   18k

Components required for the level detector

Semiconductors
D5,D6   TIL209 or similar light emitting diodes (2 off)

Integrated circuits
IC3   4011 CMOS quadruple 2-input NAND gate
IC4   4001 CMOS quadruple 2-input NOR gate

Miscellaneous
S1   push-to-make push-button, pcb mounting
Stripboard 63mm x 25mm (Vero 14354)
14-pin d.i.l. sockets (2 off)
1 mm terminal pins (3 off)
miniature crocodile clips, insulated, (1 red, 1 black)
materials or clips for probe

2 Movie show

Like the real movies, this project depends upon a characteristic of the human eye and brain known as *the persistence of vision*. A sequence of still pictures is 'projected' on to a screen in rapid succession. The pictures differ slightly from each other and the brain interprets the succession of still pictures as a moving picture. In this project the pictures are shadows cast by low-voltage lamps. There are four lamps and four pictures to the sequence, which is repeated continually. This gives a simple but very realistic repeat-action movie. We include two sequences of pictures as examples and later explain how you can draw your own movies. Artistic ability is *not* required - simple pin-people in action make amusing and effective subjects.

How it works

The circuit is driven by a clock made from two NAND gates (IC1, Fig 2.1). The rate is adjustable between 6 Hz and 20 Hz. A suitable rate for viewing is 16 Hz, which eliminates flickering. The clock pulses go to a counter (IC2) which has eight outputs, though only the first 4 (0-3) are used in this circuit. The outputs are normally low but, as pulses are fed to the counter, the outputs go high *one at a time*, in sequence. The fifth output (output 4) is fed to the reset input so that the counter is immediately reset at the 5th count and the first output (output 0) goes high.

The outputs are connected to buffers (IC3) which are non-inverting (= TRUE) gates. The buffers turn on the transistors in sequence. As each transistor is turned on, the lamp connected to it comes on. The lamps are rated at 0.3 A so they produce enough light to operate the movie show in a dimly-lit room.

How it works

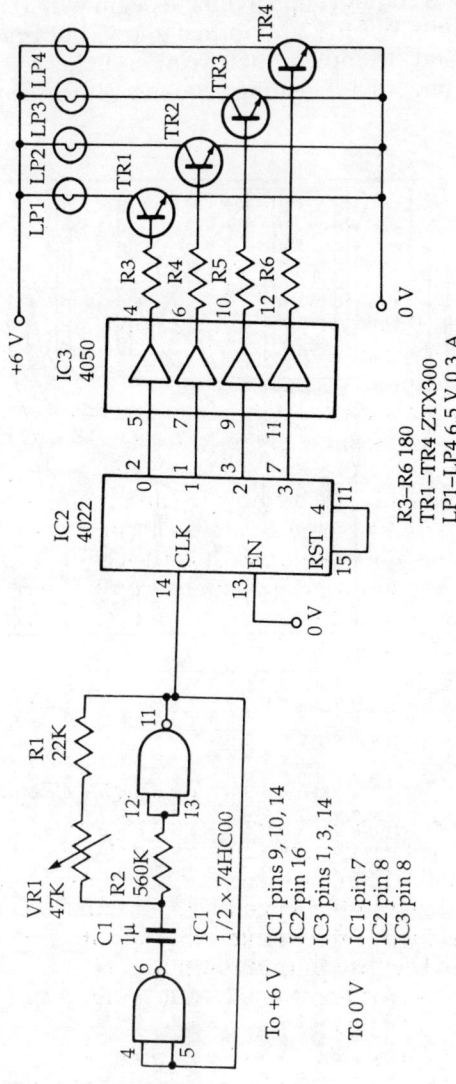

Figure 2.1 Circuit of movie show

Movie show

Construction

Assemble the clock and counter sub-circuits to begin with (Fig 2.2). Check that outputs 0 to 3 of IC2 are normally low, but briefly go high 3-4 times a second. Complete the circuit and check that the lamps flash one at a time in a repeating sequence. If the sequence

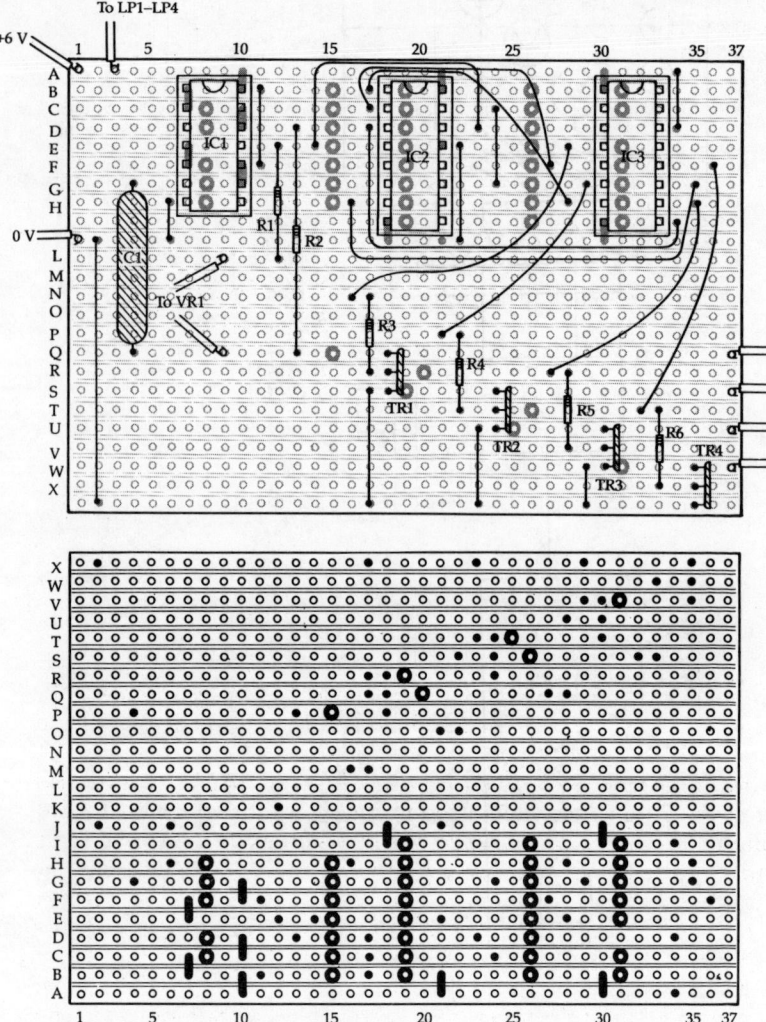

Figure 2.2 Stripboard layout of movie show

Construction

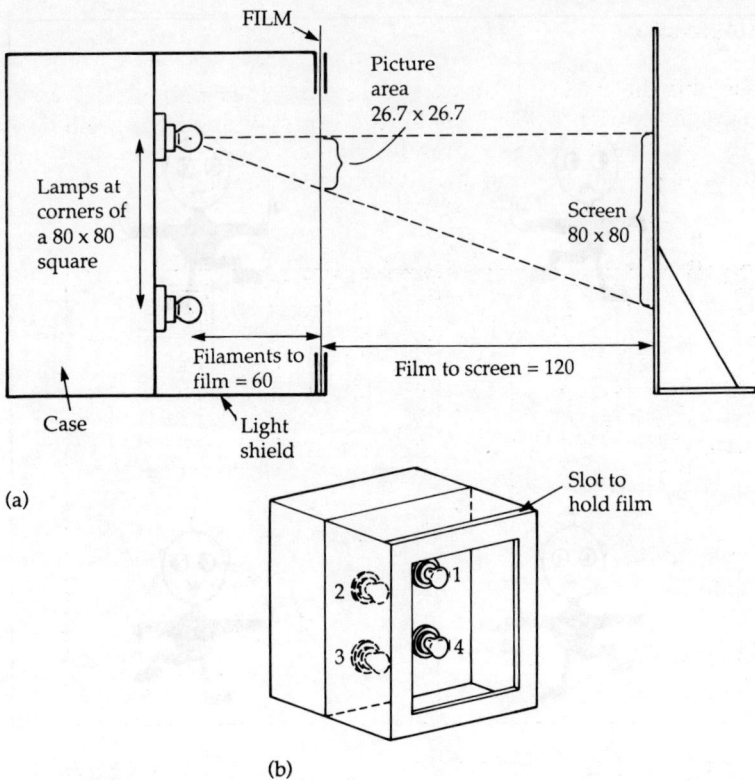

Figure 2.3 Movie show; (a) Section, (b) Showing layout of lamps. All dimensions in millimetres

appears to be wrong or any of the lamps fail to light, check the wiring on the lower half of the board. Particularly check that the copper strips have been cut at the correct places.

The project needs a plastic case to hold the circuit board and battery. Owing to the power requirements of the lamps it is more economic to use four 'D' type cells in a battery box. A case large enough to enclose the battery holder is of suitable size for mounting the array of lamps (Fig 2.3). There are two ways of mounting the lamps. The more satisfactory but more expensive method is to bolt four lamp-holders on top of the case. Alternatively, drill four 1 cm diameter holes in the top of the case, wedge the bases of the lamps in these holes and solder wires to the bases.

Movie show

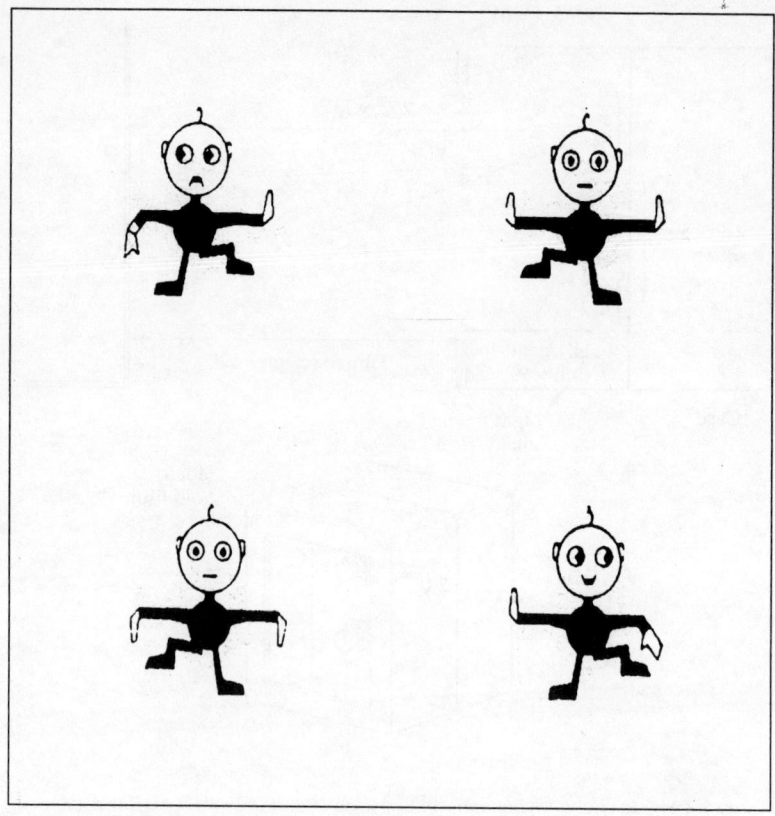

Figure 2.4 Dancing gremlin film

The light shield and film holder can be made of thin card, sheet metal or plywood. It is important that the key dimensions shown in Fig 2.3 are adhered to, otherwise the shadow images may fail to register properly when projected.

Figs 2.4 and 2.5 show two examples of 'films'. The easiest way to prepare the 'films' is to photocopy these drawings on to transparent film. Alternatively, trace them on to transparent acetate ('shirt-box') film or draughtsman's film, using a fine marker

Construction

Figure 2.5 Greedy fish film

pen or Indian ink. Fig 2.6 is a blank for use when preparing your own films. Your drawings must be aligned with the frames. These drawings will register when projected. Copyright on Figs 2.4 to 2.6 only is waived to allow you to use a photocopier for preparing the films.

When you are drawing your own films, remember that fussy detail is lost during projection. Keep the drawings simple and bold.

Movie show

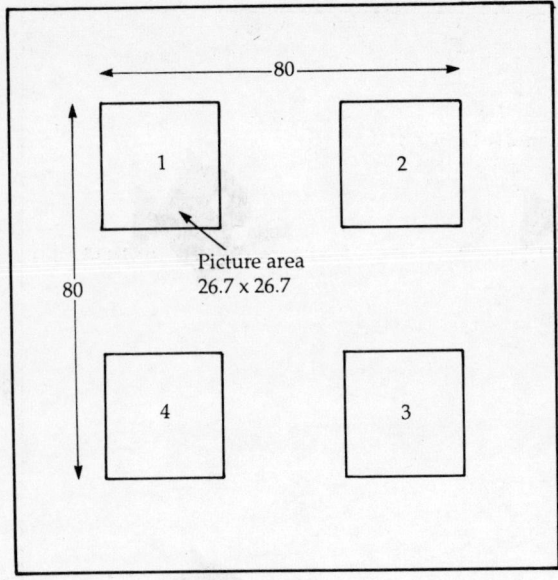

Figure 2.6 Arrangement of pictures on a film. Overall dimensions of film depend on the dimensions of the light shield

Components required

Resistors (see p.8)
R1   22 k
R2   560 k
R3–R6   180 (4 off)
VR1   47 k (or 50 k) carbon potentiometer

Capacitor
C1   1μ polyester

Semiconductors
TR1–TR4 ZTX300 npn transistor (4 off)

Integrated circuits
IC1   74HC00 CMOS quadruple 2-input NAND gate
IC2   4022BE CMOS divide-by-8 counter with 1-of-8 outputs
IC3   4050BE CMOS hex non-inverting buffer

Miscellaneous
S1   single-pole single-throw toggle or slide switch

Components required

LP1-LP4 MES filament lamps, 6.5 V, 0.3 A (4 off)
batten-mounting sockets for lamps (4 off, optional)
14-way d.i.l. socket
16-way d.i.l. sockets (2 off)
stripboard 95 mm x 63 mm (Vero 10346)
1 mm terminal pins (8 off)
knob for VR1
battery box, 4 'D'-type cells
materials for making films, light shield, film support, and screen.

3 Progressive timer

This timer is set to run for a fixed period, at the end of which it sounds an alarm for 1 second. The timer has two timing ranges: *long range* comprises periods of 5, 10, 15, 20, 25 or 30 minutes; *short range* comprises periods of 1.25, 2.5, 3.75, 5, 6.75 and 7.5 minutes. The timer has a row of light-emitting diodes (LEDs) which light up one at a time as the timing progresses (this is why it is called a progressive timer). For example, if you are timing a period of 25 minutes, the first LED lights up as soon as timing begins. After 5 minutes, this LED goes out and the next LED comes on, indicating that you are in the second 5-minute period of the 25 minutes. Every 5 minutes another LED comes on until, during the last 5 minutes, the 5th LED is lit, warning you that time is nearly up. At 25 minutes, the audible alarm sounds, and the cycle is repeated automatically.

A timer such as this is useful for timing processes that have several stages, or in any situation when you want to know how much time you have left. On its short range it is useful for timing telephone calls, since it gives you warning when you are about to begin another whole charge period of telephone time.

As we shall explain, the circuit is adaptable. The two timing ranges can be replaced by other ranges, and it is easy to provide more than just 2 ranges. The alarm can be made to sound for longer and, if preferred, can be replaced by a lamp or some other device.

How it works

This circuit makes use of a CMOS ic which, strictly speaking, is not a logic device, though its output is compatible with logic circuits. It

How it works

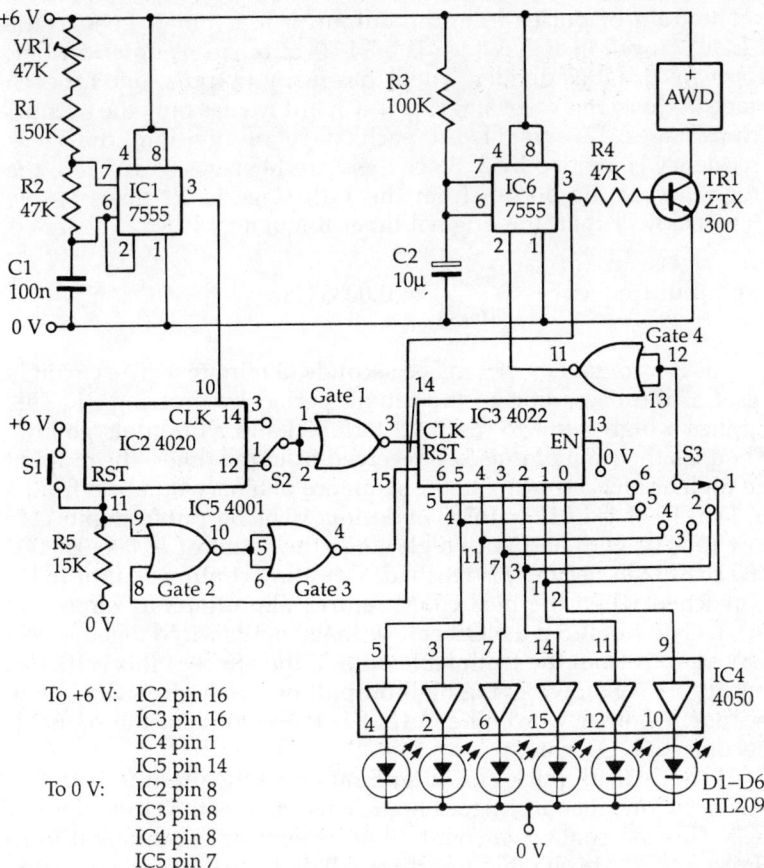

Figure 3.1 Progressive timer - circuit diagram

is a timer ic (7555) which can be used in two different ways. It can be wired either as an *astable*, to produce a continuous train of pulses at a fixed frequency, or as a *monostable*, to produce a single high pulse of fixed length. In this circuit we have both applications of the timer. In Fig 3.1, IC1 is wired as an astable. By selecting the appropriate values for R1, R2 and C1, we make it oscillate at the required frequency. The variable resistor VR1 allows the frequency to be set precisely. With the values shown, the astable can be set to oscillate at 54.61Hz. Why this particular frequency is chosen is explained in a moment, as is the function of IC6, another timer ic wired as a monostable.

Progressive timer

The train of pulses from the output of IC1 (pin 3) goes to the CLOCK input of IC2, which is a 14-stage binary counter/divider. Here we use it as divider. The ic has many outputs, one for each stage of division except stages 2 and 3, but we use only the outputs from stages 12 and 14. At each stage of division, the clock frequency is divided by 2. Since there are 14 stages of division, the frequency of the output from the 14th stage is 2^{14} times (16384 times) slower than the original timer frequency:

$$\text{Output frequency} = \frac{54.61}{16834} = 0.00333\,\text{Hz}$$

This is a frequency of 1 in 300 seconds (5 minutes). The circuit is reset at the beginning of the timing period by pressing S1. This applies a high input to the reset terminal and all outputs go low. Then, as the train of pulses is received from the timer, the outputs go high or low according to the sequence of binary numbers from 0 to 11 111 111 111 111 (=16383 in decimal). The output from pin 14 is low to start with and goes high when the count of 10 000 000 000 000 (=8192 in decimal) is reached. Counting continues until 16383 is reached when the next count returns all outputs to zeroes (all low). Gate 1 of IC5 is a NOR gate with its inputs wired together. As can be seen from the truth table on p.3, the effect of this is that its inputs are *both* low, giving high output, or *both* high, giving a low output. In other words the output is the inverse of the input. It becomes a NOT gate.

At the 16384th pulse (i.e. after 5 minutes) the input to gate 1 of IC5 goes low, its output goes high. This triggers the clock input of IC3. This too contains a counter but its outputs are different from those of IC2. This has 8 outputs, numbered 0 to 7 (we use only 0 to 6), At each stage *one* of the outputs is high and the others are low. To begin with, output 0 is high but, when the clock input changes from low to high, output 0 becomes low and output 1 becomes high. Every 5 minutes, as each 16384th pulse is received by IC2, the counter of IC3 is advanced by one step. In turn, outputs 1, 2, 3, 4 and so on, go high. The state of these outputs is shown by a row of LEDs, D1 to D6. These are powered from IC4, which has 6 buffer gates. These are logic TRUE gates, since the output has the same logic state as the inputs. Thus, when the timer is running, the LEDs light up in order, one at a time, indicating the progress of the timing period.

The selection of short or long range is made by S2 which connects Gate 1 either to stage 14 or stage 12 of IC2. Stage 12 represents division by 2^{12} (=4096 decimal), so its frequency is 1 in 1.25 minutes instead of 1 in 5 minutes.

The selection of the total timing period is made by setting the 6-way rotary switch S3. To set the timer for 15 minutes, for example, S3 is turned to position 3. At the end of 15 minutes, output 2 of IC3 goes low and output 3 goes high. Until this point, both inputs of gate 4 of IC7 have been low and its output has been high. As soon as output 3 goes high, a low level appears at the output of gate 4. This triggers the monostable, IC6. The values of R3 and C2 determine that its high pulse lasts 1.1s. As it receives the low triggering pulse from gate 4, its output (pin 3) goes high. TR1 is turned on for 1.1s and a current flows through the audible warning device (AWD). This is a semiconductor 'buzzer' or sounder that produces a note when current passes through it.

The high output from IC6 also goes to gate 2 of IC5. Previously both its inputs have been low, so its output has been high. Gate 3 is wired as a NOT (or INVERT) gate, so the input to the reset terminal of IC3 has been low. Now that one input of gate 2 has been made high, the reset input of IC3 becomes high. This instantly resets the counter of IC3. Thus, as output 3 goes low, output 4 goes high for a few microseconds, the counter is reset and output becomes high. the sequence then repeats. If S3 is set to position 3 as already described, the AWD bleeps regularly every 15 minutes. Thus the timer can be used to produce bleeps at fixed intervals ranging from 1.25 to 30 minutes.

The circuit can be reset at any time during the sequence described above. Pressing S1 causes a high level at the reset of IC2, resetting the counter. It also makes the output of gate 2 go low, and causes a high level at the reset of IC3. The first LED comes on and timing recommences.

Construction

Fig 3.2(a) shows how the circuit is wired up on a piece of stripboard. Begin by assembling the timer and first counter circuit (IC1, IC2, VR1, R1, R2 and C1) including part of the reset sub-circuit (S1, R5). This may now be tested. The input to pin 11 of IC2 is normally low, but goes high when S1 is pressed. The output from stage 12 (pin 1) changes from low to high and back again every 1.25 minutes (75s). Monitor this with a voltmeter or the logic probe (Project 1). For preliminary checking, it is quicker to monitor one of the earlier stages, such as stage 10 (pin 14) which changes

Progressive timer

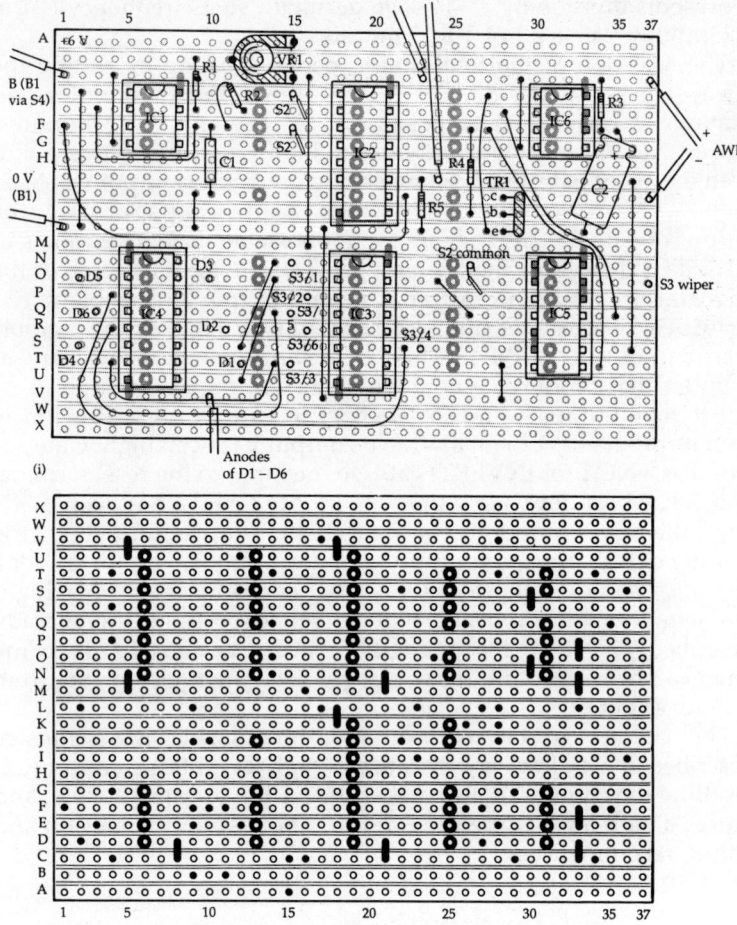

Figure 3.2 Progressive timer stripboard layout; (a) main board

from high to low and back again approximately every 20 seconds. Adjust VR1 to get this timing right, then monitor the output from pin 1 while you make the final adjustments.

Next wire up IC3-IC6, together with S2, S3, R3, C2 and the LEDs. Fig 3.2(b) shows how the LEDs may be mounted on a small scrap of strip-board. Fig 3.3 shows the off-board wiring. Switch S2 to short range (IC2, pin1) and S3 to position 6. Check that the '0'

Construction

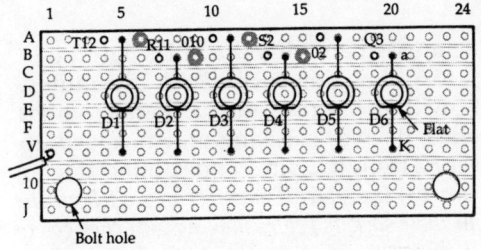

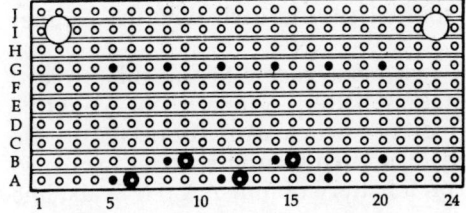

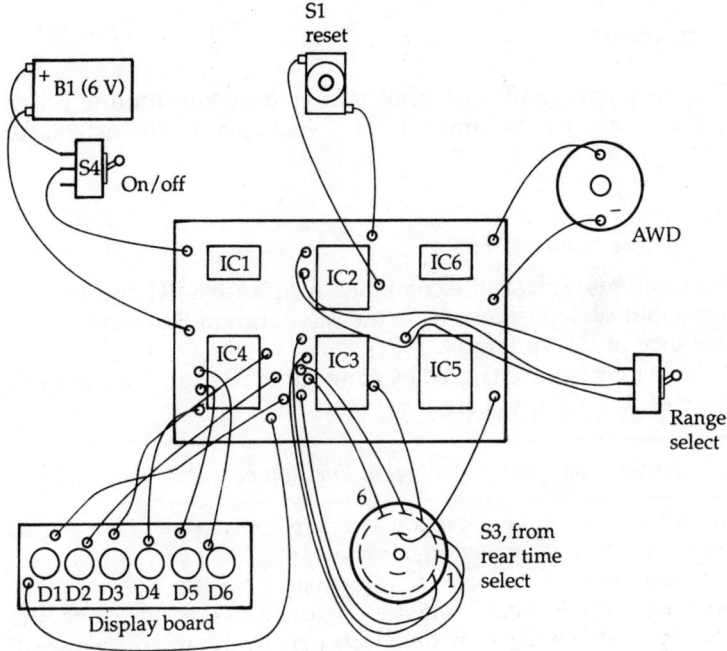

Figure 3.3 Progressive timer, off-board connections

29

LED comes on when reset is pressed and that the LEDs light up in a repeating sequence from 1 to 6, changing once evey 1.25 minutes. If there is something wrong with the wiring and LEDs fail to light or light in the wrong sequence, it saves checking time to speed up the clocking rate. Disconnect S2 from IC2 pin 1 and connect it to an earlier stage (e.g. stage 8, pin 13) instead. *Remember to switch off the power, and remove the ics from their sockets before making any changes to the circuit.*

Finally add the AWD sub-circuit, R4, TR1 and the AWD. Check that the AWD sounds for about 1 second each time the counter circuit resets.

A variety of small plastic cases is available to suit this project. One of those with an integral 6 V battery compartment would be the most suitable. A power on/off switch S4 is connected in the positive lead from the battery. The circuit requires about 13 mA, so it runs for about 150 hours on a battery of four AA (HP7) alkaline cells.

Modifications

(1) The basic timing of the circuit is altered by substituting other values for VR1, R1, R2 and C1. The frequency of the astable is calculated from:

$$f = \frac{1.44}{(R_A + 2.R_B)C}$$

where f is in hertz, R_A is the combined resistance of R1 and VR1 set to some mid-way position, R_B is the resistance of R2, and C is the capacitance of C1, in farads.

(2) Time ranges are changed by connecting S2 to any two or more outputs of IC2. Outputs are:

Stage	Division by	Pin	Stage	Division by	Pin
1	2	9	9	512	12
4	16	7	10	1024	14
5	32	5	11	2048	15
6	64	4	12	4096	1
7	128	6	13	8192	2
8	256	13	14	16384	3

(3) The period for which the alarm sounds is altered by substituting other values for R3 and C2: The 'on' period of the monostable is calculated from:

t = 1.1RC

where t is the time in seconds, R is the value of R3 and C is the value of C2, in farads. This period must not be longer than 1.25 minutes, for this holds IC3 in the reset condition when it is due to accept the first count.

(4) Other devices may be substituted for the AWD, such as an LED, a small filament lamp (6 V, 0.06A), a solid state siren, or a miniature 6 V relay. The relay can be used to switch many other devices, such as an electric bell or motor, which may be mains-powered if a suitable relay is chosen. *Beginners are advised not to attempt wiring a mains-powered device to the circuit* without seeking the advice of an experienced person.

Components required

Resistors (see p.8)
R1   150k
R2, R4   47k (2 off)
R3   100k
R5   15k

Capacitors
C1   100n polycarbonate layer
C2   10µ electrolytic

Semiconductors
D1-D6   TIL209 or similar light-emitting diodes (6 off)
TR1   ZTX300 npn transistor

Integrated circuits
IC1, IC6   7555 CMOS timer (2 off)
IC2   4020BE 14-stage binary counter/divider
IC3   4022BE divide-by-8 counter with 1-of-8 outputs
IC4   4050BE hex non-inverting buffer
IC5   4001BE quadruple 2-input NOR gate

Miscellaneous
S1   push-to-make push-button
S2   single pole double throw switch
S3   6-way rotary switch

Progressive timer

S4   single-pole single-throw toggle or slider switch, low voltage rating
stripboard 95 mm x 63 mm (Vero 10346), 63 mm x 25 mm (Vero 14354)
1 mm terminal pins (21 off)
suitable plastic enclosure
8-way d.i.l. sockets (2 off)
14-way d.i.l. socket
16-way d.i.l. sockets (3 off)
knob for S3
battery clip or 4-cell battery holder (if integral battery compartment not provided)
nuts, bolts and washers for mounting circuit boards (or use self-adhesive mounting strip), connecting wire, solder.

4 Intruder detector

This detector is based on the principle of having a transmitter to produce a beam of radiation and a receiver to detect it. If an intruder comes between the transmitter and receiver, the beam is broken and the alarm is sounded. We use a beam of infra-red radiation, which is invisible to the eye. This makes it impossible for the intruder to notice the beam and avoid breaking it. With some systems it is possible for the intruder to saturate the detector by shining a bright light at it while stepping through the beam. This can not be done with this receiver, as it is tuned to accept only a pulsed beam of fixed frequency. Indeed, if the intruder flashes a torch around the room and accidentally shines it in the 'eye' of the receiver, this alone may be sufficient to trigger the alarm.

The distance between transmitter and receiver can be up to 2m, possibly more. The receiver circuit is able to trigger a wide range of devices.

How it works

The transmitter circuit (Fig 4.1) is based on an astable multivibrator running at about 2 kHz. The period for which the output is high is about 500μs, but the period for which it is low is only 8μs. The waveform of the astable is inverted by the transistor TR1, so the field-effect transistor TR2 is switched on for 8μs in every 500μs. In other words, its *duty cycle* is 1.6%. When TR2 is switched on, a large current flows through the infra-red diodes D1 and D2. The current is about 1.5 A through each diode but, since they are on for only 1.6% of the time, the *average* current through each diode is only about 24 mA. This means that the LEDs emit pulses of

Intruder detector

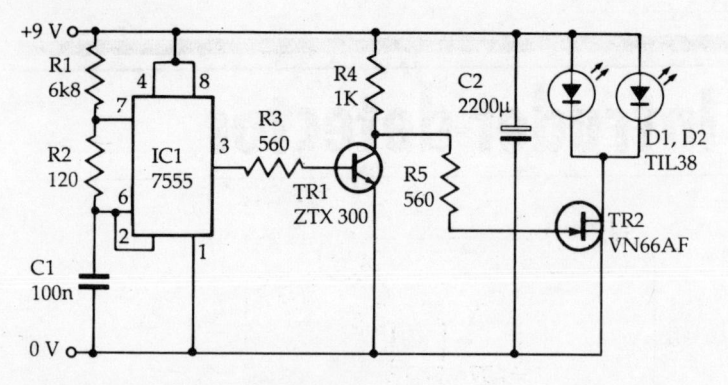

Figure 4.1 Transmitter for intruder detector

radiation of high intensity while requiring only a low average current. The low average current is not sufficient to burn out the LEDs. The high intensity gives a good working range to the device, without requiring high amplification at the receiver.

The receiver (Fig 4.2) has only one stage of amplification, the operational amplifier IC2. This is wired as a comparator. The (−) input comes from the positive end of R6. The infra-red photodiode D2 is reverse biassed, and the amount of leakage current through it depends on the amount of infra-red it is receiving. As the leakage current varies so does the voltage across R6. The (−) input therefore is fairly close to 0 V, rising when a brief pulse of infra-red arrives. The voltage at the (+) input of the amplifier is set by VR1, to lie between the lower and upper voltage levels of the (−) input. There is no feedback resistor on the amplifier so its output swings sharply from high (6 V) to low (0 V) and back at each infra-red pulse.

The next stage is a phase-locked loop, the circuit of which is contained in IC3. The function of the PLL is to detect and lock on to an incoming train of pulses that has a frequency within the range to which it is tuned. The PLL is tuned (by means of C3, R9 and R10) to respond to a frequency lying between 1 kHz and 3 kHz. The frequency of the transmitter is exactly in the middle of this range. If the frequency being received is within this range, a logic high output appears at pin 10. Otherwise a logic low appears. Thus, as long as the beam is not interrupted, output from pin 10 is high. Interruption of the beam, or saturation of D3 with a strong *continuous* source of infra-red, makes output go low.

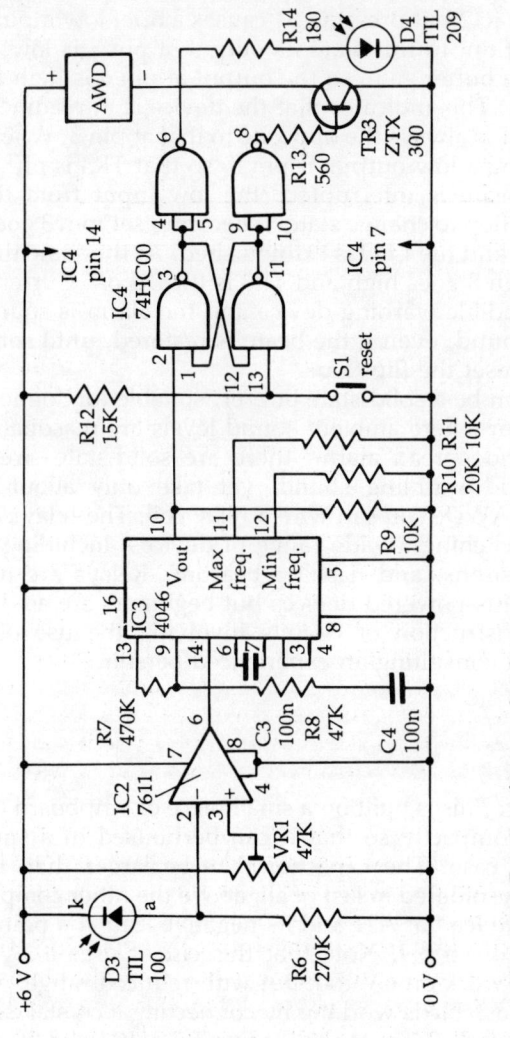

Figure 4.2 Receiver for intruder detector

A low output from IC3 is detected by the flip-flop consisting of two cross-coupled gates of IC4 (see also Fig 1.2). Two further gates are used as buffers to drive an LED and a transistor respectively. Normally the inputs to the flip-flop (pins 2 and 13) are high. Pin 13 is held high by R12 but pressing S1 causes a brief low input which resets the flip-flop. In this state its output at pin 3 is low. This is inverted by the buffer gate, so the output at pin 6 is high and the LED (D4) is on. This indicates that the device is in readiness. The output at pin 11 is always the opposite to that at pin 3. When reset, it is high, giving a low output at pin 8, so that TR3 is off.

When the beam is interrupted, the low input from the PLL causes the flip-flop to change state. It becomes set; pin 3 goes high, pin 6 goes low and the LED is extinguished. At the same time, pin 11 goes low, pin 8 goes high and TR3 is turned on. Current flows through the audible warning device and the alarm is sounded. It continues to sound, even if the beam is restored, until somebody presses S1 to reset the flip-flop.

The AWD can be a solid-state buzzer, suitable for sounding the alarm in a room where ambient sound levels are reasonably low. For a more widespread alarm, there are solid-state sirens that generate a loud warbling sound, yet take only about 20 mA. Instead of the AWD, you can wire a relay coil. The relay can then be used for switching a wide range of devices, including lamps, electric bells, sirens, and flashing beacons. Relays are used for controlling mains-powered devices but beginners are advised not to attempt construction of circuits involving the use of mains power without consulting an experienced person.

Construction

(1) Transmitter: This is built on a small piece of stripboard (Fig 4.3) and laid out compactly so that it can be housed in a small and inconspicuous case. The capacitor may be larger than the one illustrated. It is soldered in last of all, above the other components, with its positive lead at A21 and its negative lead at a point to the left-hand end of strip J. Note that this circuit uses a 9 V power supply. It will work on 6 V too, but with reduced range. You can check that the astable is working by connecting a crystal earphone in series with a 100n capacitor between pin 3 of IC1 and the 0 V rail. A high-pitched continuous note is heard. To test the completed circuit, connect a milliammeter in series with the power supply. The circuit takes about 50-60 mA. If only a small current is flowing,

Construction

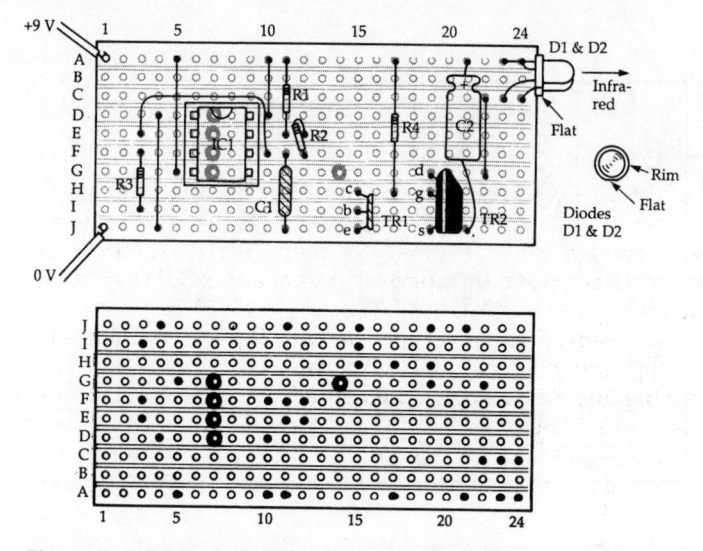

Figure 4.3 Infra-red transmitter for intruder detector

check that the diodes have been connected the right way round. The 'flat' on the rim of the diode is toward the bottom of the drawing.

A battery could be used for power but, since the transmitter is likely to be switched on for periods of many hours, it is more economical to use a mains adaptor ('battery eliminator'). Units giving 9 V dc at up to 300 mA are inexpensive and very suitable.

(2) Receiver: This is also set out very compactly to minimise the size of the case required for it. The infra-red diode is reverse-biassed so its *cathode* is wired to the positive rail. The diode is oriented to receive radiation from the left in Fig 4.4, its chamfered corner being toward the bottom of the drawing. There is no cut in the strip joining pin 3 to pin 12 of IC4. When power is applied, the voltage at the (−) input of IC2 is less than 1 V. Turn VR1 so that the voltage at its wiper is about 1.5 V. The output at pin 6 of IC2 is 5 V or more. The PLL is tested by placing the transmitter close to the receiver with D1 and D2 directed at D3. Temporarily wire the terminal pin at L17 to the 0 V rail. The effect of this is that D4 indicates the output of IC3. Apply power to both circuits, and D4 lights. Place your hand to break the beam and D4 goes out. Try moving the circuits further apart; you may need to re-adjust VR1 to obtain the correct action. Beyond 20-30 cm the receiver ceases to

37

Intruder detector

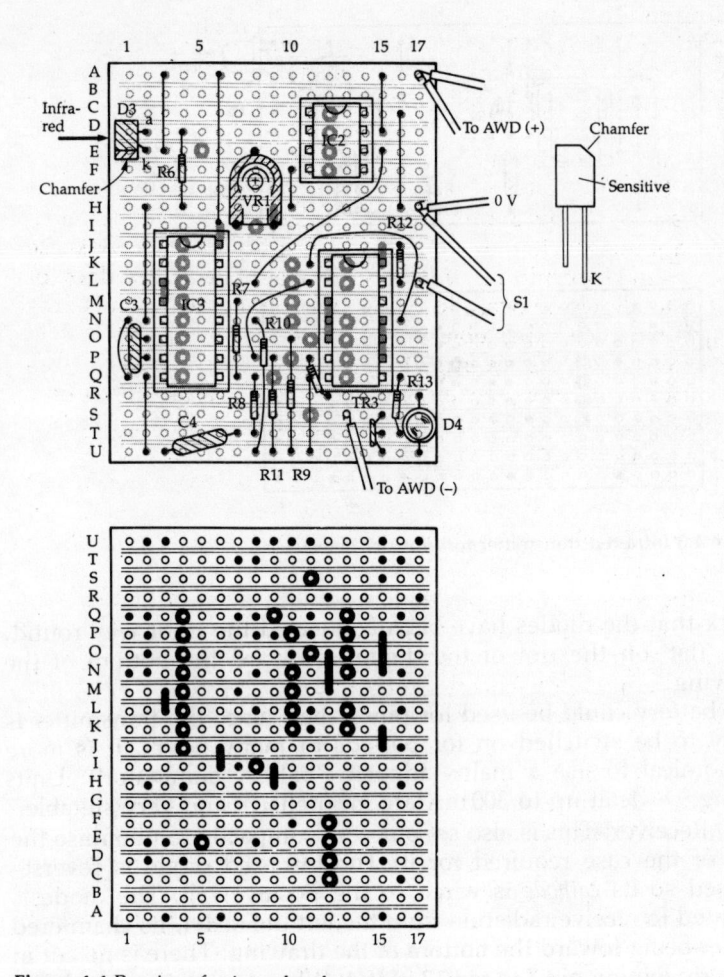

Figure 4.4 Receiver for intruder detector

function. It needs a lens to concentrate the radiation on D3, as explained in the next section.

The receiver is best powered by a 6 V mains adaptor though, as it takes only 17.5 mA (or only 7.5 mA if you omit D4) it is feasible to use a 6 V battery.

Installing the circuits

The circuit-boards may be enclosed in small cases mounted on the wall or resting on furniture. It is usually important for the cases to be inconspicuous. This is less important in protecting a doorway that can only be approached by the intruder from one direction. If the cases are placed on the far side of the doorway, the intruder will have broken the beam before having spotted the cases. It is often possible to hide cases partly behind window curtains. Another approach is to hide a circuit inside a cupboard, a bookcase or other item of furniture, with a small hole cut in the door or panelling to allow the beam to pass. You could even disguise the circuits as books or other objects.

The receiver can be swamped by an excess of infra-red from other sources. Therefore ensure that there is no other strong source in the receiver's field of view. If the system is to be used in daylight, check that direct sunlight can not fall on the receiver.

As explained above, the receiver requires a lens. Special red plastic lenses are available. These usually have a focal length of about 8 cm for infra-red, so should be mounted 8 cm in front of the receptive surface of D3. It is also possible to use any small magnifying lens; a cheap plastic lens about 3-5 cm in diameter is adequate. Choose one with a focal length of 4-10 cm. You can find the focal length by focussing an image of the Sun or a distant lamp on to a piece of paper. This is the focal length for visible light; the focal length for infra-red radiation is slightly greater. However, you do not need to know the precise focal length since all that is necessary is to concentrate the beam on to an area about 3 mm in diameter on the surface of D3. It does not have to be focussed to a sharp point. Mount the lens behind a hole of suitable diameter cut in the receiver case. It is best if the interior of the case is black to eliminate possible interference from other sources of infra-red. Bore a small hole in the case so that you can adjust VR1 with a screwdriver, and another small hole so that you can observe D4.

At the extreme of the range (beyond 2m) the transmitter and receiver need careful optical alignment. It is better to mount the transmitter in its final position and to begin with the receiver closer than eventually intended. D4 lights, but goes out when the beam is broken. Gradually move the transmitter away from the receiver, watching D4. Adjust VR1 if it goes out. You may also need to adjust the alignment of the diodes by bending their leads slightly. The PLL takes a few cycles to lock on to the pulse train, so make all adjustments slowly. Otherwise you may over-adjust before the PLL has time to respond. Once the system has been optically aligned, make sure that it is firmly secured in position.

Components required

Resistors (see p.8)

R1	6k8	R7	470k
R2	120	R8	47k
R3,R5,R13	560 (3 off)	R9,R11	10k (2 off)
R4	1k	R10	20k
R6	270k	R12	15k
		R14	180

VR1   47k miniature horizontal preset potentiometer

Capacitors
C1,C3,C4   100n polyester (3 off)
C2   2200µF electrolytic (or 1000µF)

Semiconductors
D1,D2   TIL38 high-power infra-red LED (2 off)
D3   TIL100 infra-red photodiode
D4   TIL209 or similar red LED
TR1,TR3   ZTX300 npn transistor (2 off)
TR2   VN66AF VMOS n-channel enhancement mode FET

Integrated circuits
IC1   7555 CMOS timer
IC2   7611 CMOS operational amplifier
IC3   4046B CMOS phase-locked loop
IC4   74HC00 CMOS quadruple 2-input NAND gate

Miscellaneous
stripboard 63 mm x 25 mm (Vero 14354), and 45 mm x 56 mm
1 mm terminal pins (5 off)
suitable plastic enclosures
lens
8-way d.i.l. sockets (2 off)
14-way d.i.l. socket
16-way d.i.l. socket
audible warning device (solid state buzzer, solid state siren, etc)
6 V dc and 9 V dc mains adaptors, 300 mA unregulated

5 Capacitance meter

A capacitance meter has innumerable uses as a test instrument for the electronic workshop. In particular, it is useful for checking the capacitance of electrolytic capacitors, since their actual capacitance may differ by as much as 50% from the value printed on the can. Another problem is that the markings on capacitors so often seem to have been printed in an ink specially formulated to smudge easily, leaving an indecipherable blur. Using a capacitance meter avoids the consequent uncertainties. A ready-made meter is usually expensive and, even in kit form, can be a major item. This design cuts costs and reduces the complexity of the wiring by using a much simpler display. It may be a little less convenient to use than the proprietary instruments or kits, but is much easier for a beginner to construct.

How it works

The circuit is based on two timer ics and, as in Project 3, uses one in the astable mode and one in the monostable mode. The astable is used to generate a series of pulses at a known high frequency. The monostable produces a single high pulse of length determined by the test capacitor. The logic counts how many pulses are produced by the astable during the single pulse of the monostable. The number of pulses depends upon the capacitance of the test capacitor, which can then be read off from a table.

In Fig 5.1, IC1 is the monostable and IC2 is the astable. Resistors and a capacitor of 1% tolerance are used with IC2, their values giving a frequency of 5958±60 Hz. The period of the monostable, which also has a 1% tolerance resistor, is 11s if R2 (10 k) is switched

Capacitance meter

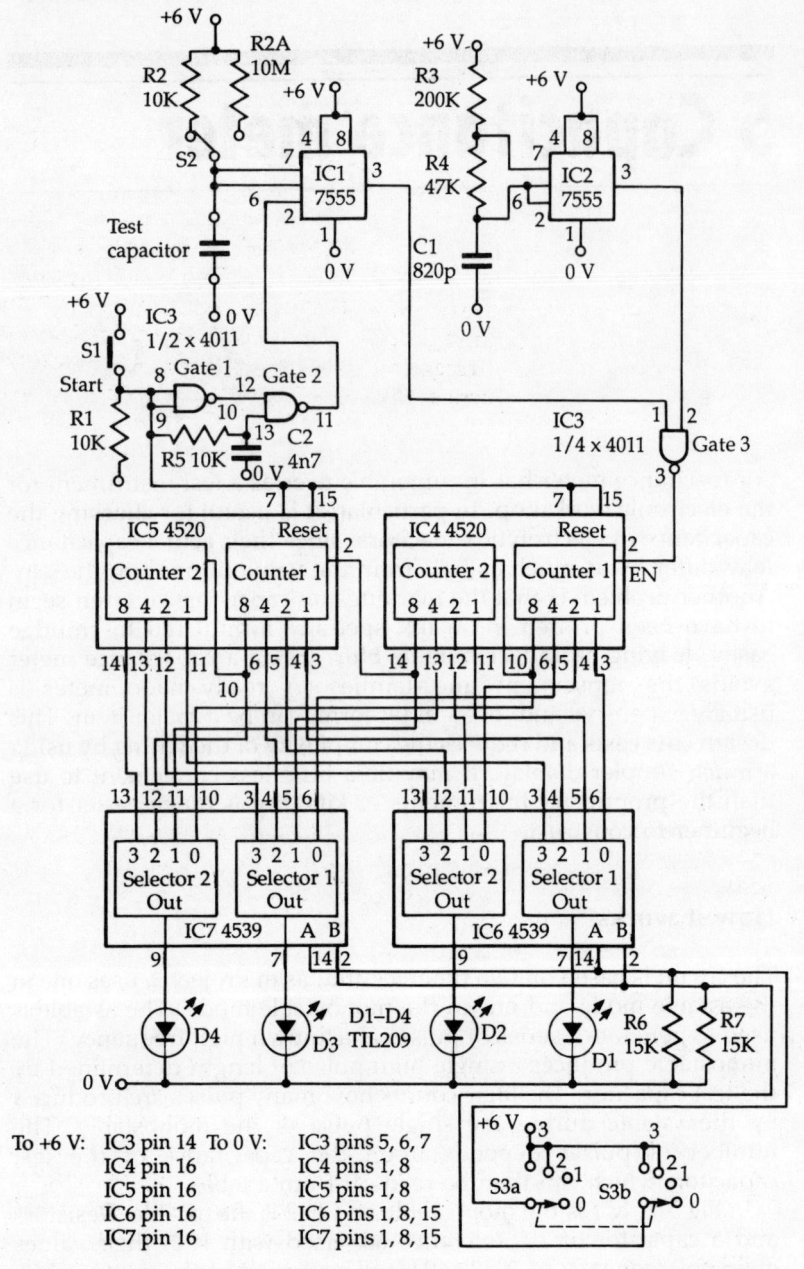

Figure 5.1 Circuit of capacitance meter

into circuit and the test capacitance is 1000 µF, and proportionately shorter if it is less. If R2A (10 M) is switched in, the period is 11s with a 1 µF test capacitor. We switch in R2 for measuring capacitance in the range 100 nF to 1000 µF, and use R2A for the range 100 pF to 1 µF. The measurement begins when S1 is pressed and released. This sends a high pulse to the reset inputs of the counters (IC4,5), making their outputs all low. Gates 1 and 2 of IC3 are wired as a pulse-generator, to give an exceedingly short high pulse as S1 is released. This triggers IC1 into action. While the output of IC1 is high, pulses arriving from the astable (IC2) at gate 3 of IC3 are passed through to the enable input of IC4. IC4 and IC5 each contain two 4-bit binary counters. These are wired in cascade to give a 16-bit binary counter. Counting continues for as long as the output of IC1 is high, a period which depends on the value of the test capacitor. If the test capacitance is 1000 µF, the counting period lasts 11s and the number of pulses counted is 65536. This is 2^{16}, or 1 followed by 16 zeroes in binary. A 16-bit counter can register up to 65535, so a 'full' counter corresponds to 1000 µF, except for the last pulse to be counted, which is well within the range of precision required.

Commercially produced capacitance meters usually convert the binary count to decimal and show readings on a 3-digit or 4-digit numeric display. The circuit for such conversion and display is complex and expensive so we have a much simpler binary display in this project. However, a 16-bit binary display with its row of 16 LEDs is not at all easy to read. Instead we have 4-bit display using only 4 LEDs, which can be switched to show the output state of any one of the four counters. Switching is done by data selectors in IC6 and IC7. Each IC contains two selectors. Each of the four selectors has 4 inputs, 0 to 3, but only 1 output. The output has the same state as *one* of the inputs, depending on which input is being selected at that time. An input is selected by applying logic levels to the data select inputs A and B:

Select input levels B A	Data input selected	Data comes from
0 0	0	IC4 counter 1
0 1	1	IC4 counter 2
1 0	2	IC5 counter 1
1 1	3	IC5 counter 2

A rotary switch S3 is connected so as to bring the select inputs A and B to the levels indicated in the table. As S3 is rotated, the output state of one of the counters appear on the LEDs.

The capacitance is calculated by adding together a few numbers, depending on which LEDs are lit. This table gives the capacitances corresponding to each LED for each setting of S3.

Position of S3	LED lit: D4	D3	D2	D1
0	122	61	31	15
1	2	1	488	244
2	31	16	8	4
3	500	250	125	63

With R2 in circuit (high range), values in *italics* are nanofarads; the rest are microfarads. Thus, if S3 is in position 2, and LEDS D3 and D1 are lit, the capacitance is 16+4 = 20 µF. This is assuming that it has been checked that no LEDs light when S2 is in position 3. With R2A in circuit (low range), the table values in *italics* are picofarads; the rest are in nanofarads. Thus, if S3 is in position 1 and D4 and D2 are lit, the capacitance is 2.488 nF (call it 2.5 nF), or 2500 pF. Results of the addition should always be rounded off to the nearest 1%.

A reading to this degree of accuracy is usually precise enough but, if greater precision is required, digits of lower significance can be taken into account. In the first example above, turning S3 to position 1 might reveal that D4 and D3 are lit. This corresponds to an additional capacitance of 2+1 = 3 µF, so the total capacitance is 25 µF, to the nearest 1 µF. It is possible, though hardly necessary, to obtain even finer resolution by adding in the values obtained with S3 in position 0.

Construction

Begin by building the monostable and astable. Solder a pair of wires terminating in crocodile clips to the pins at G14 and K14. The wire and clip connected to G14 should be red, while the wire and clip at K14 should be black or blue. This is important as it is essential that electrolytic and tantalum capacitors are connected with the correct polarity, red to positive and blue or black to negative. Test the monostable (IC1) by clipping a 1000 μF capacitor into the circuit and briefly connecting pin 2 to +6 V. Its output goes high for 11s, and can be measured with a voltmeter or the logic level detector (Project 1). The output from the astable (IC2) has too high a frequency to cause a visible flicker with the logic level detector; *both* LEDs glow at half brightness.

Next add IC3 to the circuit, as well as S1, R1, R5 and C2. Note that one gate of IC3 is not used and its inputs at pins 5 and 6 are connected to 0 V by means of the solder blob shown in the figure. You can use the pulse detector (Project 1) to check that pin 11 gives a high pulse when S1 is released. Or wire pin 11 to IC1 and check that it produces an 11-second high output when S1 is released. The output from pin 3 of IC3 is normally high but during that 11s, it alternates rapidly. Both LEDs on the logic level detector glow at half brightness.

The circuit is now ready to receive the counters, IC4 and IC5. At this stage do not make the connections to the selectors. The circuit may now be tested, using a test capacitor of 100 μF or 470 μF. Connect a voltmeter or the logic level probe to pin 14 of IC1. Press S1 and release it. The output changes rapidly but visibly between high and low for a few seconds and then settles to high or low. This shows that a count is being registered. Now measure and make a note of the 16 counter outputs. Calculate the corresponding capacitance. Do not expect the calculated value to be close to the value printed on the can. Repeat with a few other capacitors of larger and smaller capacitance.

Finally assemble the data selector sub-circuit, IC6, IC7, D1 to D4, S3, R6 and R7. Fig 5.2 shows which IC and which pin the connections go to; for example the connection at T4 goes to IC7, pin 10. As you solder each wire in place, check that it is correctly inserted. The LEDs may be mounted on the board as illustrated, or they may be mounted on the case and leads taken from terminal pins soldered at EE36 to HH36 to the anodes of the LEDs. A lead connecting the cathodes of the LEDs is run back to the 0 V supply. Turn S3 successively from position 0 to position 3 and, at each step check that the LEDs are lit corresponding to the outputs from counters.

If the project is to be used only occasionally, it can be left unenclosed. If it is intended for frequent use, mount the circuit board in a case, preferably one with an integral battery compartment. The circuit requires only 150 mA, so 4 pen-light cells will be sufficient. If the project is enclosed, the conversion table could be marked on the case-top, as in Fig 5.4. The crocodile clips

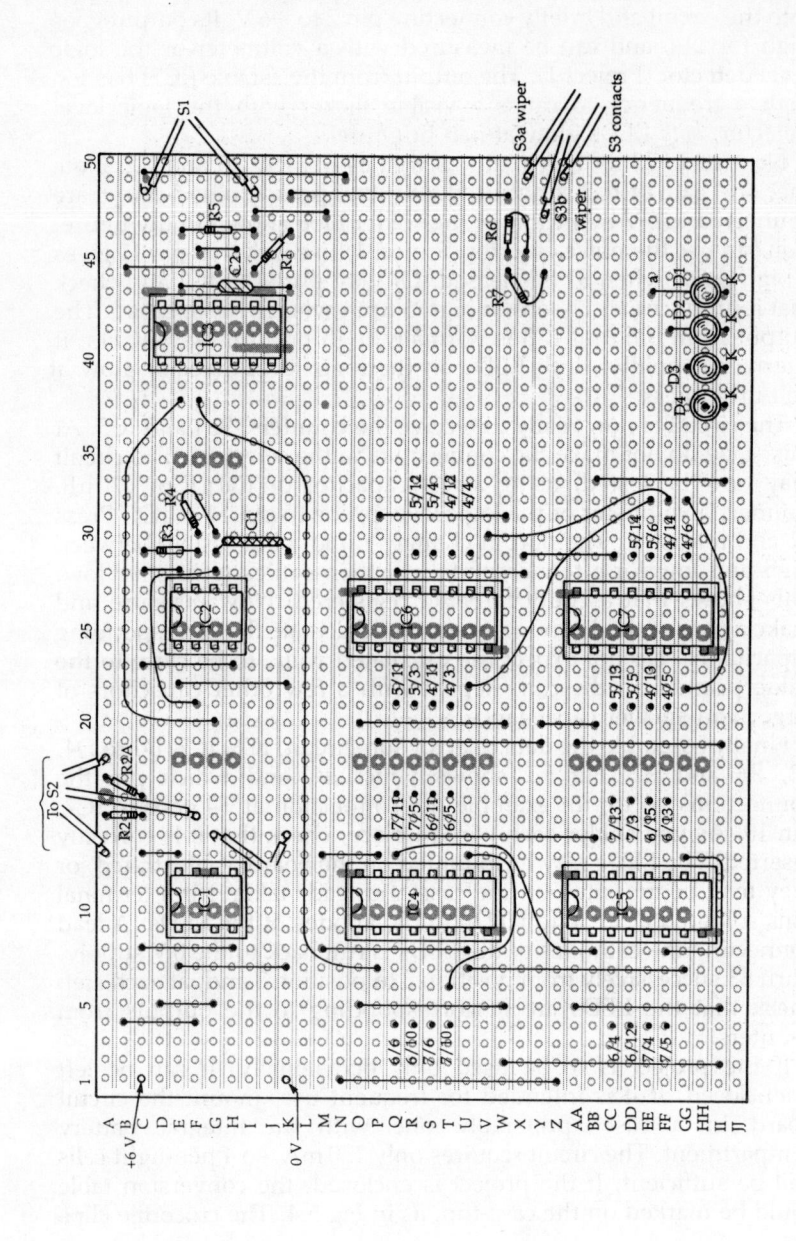

Capacitance meter

Construction

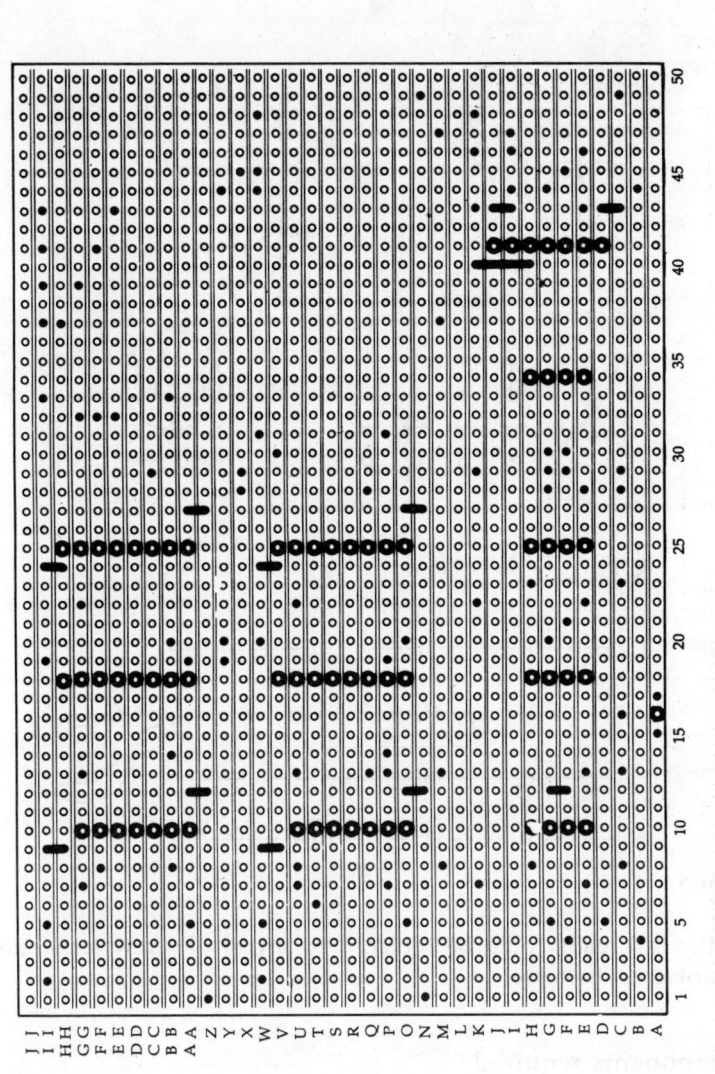

Figure 5.2 Stripboard layout of capacitance meter

Capacitance meter

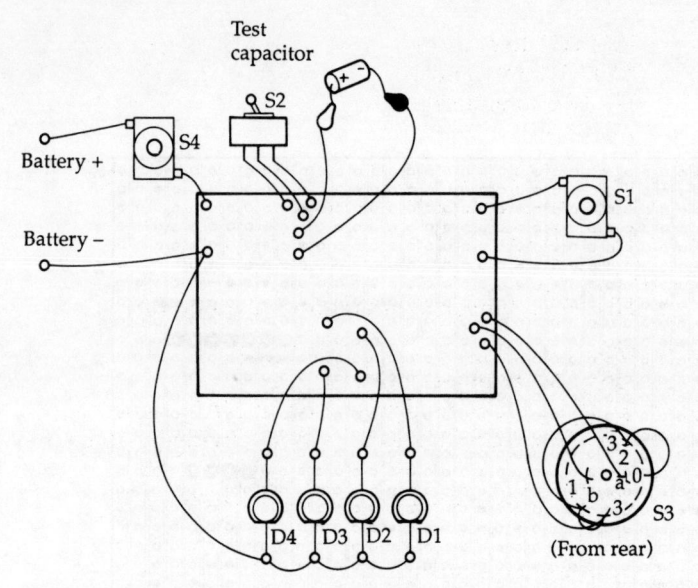

Figure 5.3 Off-board connections of capacitance meter

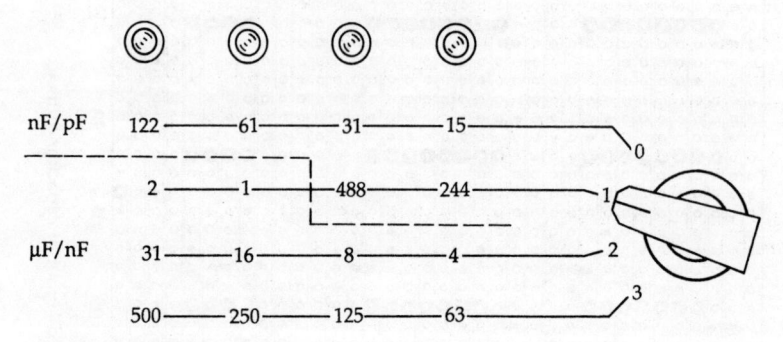

Figure 5.4 Marking the panel

on the test leads could be replaced by a pair of screw terminals mounted on the case.

Components required

Resistors (see p.8)
R1,R5   10k (2 off)
R2   10k (1% tolerance)

Components required

R2A	10M (1% tolerance)
R3	200k (1% tolerance)
R4	47k (1% tolerance)
R6,R7	15k (2 off)

Capacitors
C1	820p silver-mica, 1% tolerance
C2	4n7 polyester

Semiconductors
D1-D4	TIL209 or similar light-emitting diode (4 off)

Integrated circuits
IC1, IC2	7555 CMOS timer (2 off)
IC3	4011BE CMOS quadruple 2-input NAND gate
IC4, IC5	4520BE CMOS dual synchronous divide-by-16 counter
IC6, IC7	4539BE CMOS dual 4-input data selector

Miscellaneous
S1	push-to-make push-button
S2	single-pole double-throw toggle switch
S2	2-pole 4-way rotary switch (2-pole 6-way is suitable)

stripboard 127 mm x 95 mm (Vero 10347)
1 mm terminal pins (12 off, or 16 off if LEDs not on board.
suitable plastic enclosure (optional)
8-way d.i.l. sockets (2 off)
14-way d.i.l. socket (1 off)
16-way d.i.l. sockets (4 off)
crocodile clips or screw terminals (1red, 1 black/blue)
knob for S3

6 Combination door sentry

This is a pure logic circuit - almost an electronic puzzle, and if you study the way this works, you will learn a lot about logic gates. The sentry guards the door of a room or cupboard and, unless you key in the correct combination of numbers, you can not or dare not open the door. It can also be used to guard electrically powered machinery or equipment. As with most combination locks and with the PINs used for bank cash dispensers, you have to enter a particular set of four digits. To save the expense of a full keyboard, this circuit operates with only 3 keys. It has an LED display to show the digits that have been keyed. The sequence for opening the lock is as follows:

(1) Press key C to cancel any previous entries: the display shows '0'.

(2) Press key P a number of times, to key in the first digit. At each press, the display is incremented by 1.

(3) When the correct first digit is showing on the display, press key E to enter it - take care to press E only once.

(4) Repeat steps 2 and 3 for the other three digits.

(5) You may now open the door.

(6) If you think you have made a mistake in steps 1 to 4, press key C to cancel what you have done and begin again, from the first digit.

The sentry guards the door in one of two ways. One way is to have an electrically-operated bolt that can be opened only after the correct combination has been entered. The bolt is operated by a switch but, if you try to operate the bolt without having entered the correct combination, an alarm sounds. The other way has no special bolt, but the alarm sounds automatically if the door is opened without the proper combination having been keyed in.

How it works

The P ('press') and E ('enter') keys are used to operate two counters (IC2) which count the number of times each key is pressed. Ordinarily when a key is pressed, its contacts make and break several times before they finally make. We call this *contact bounce*. The result is that a single key-press produces a series of high and low pulses. In this circuit a single key-press would advance the counters by several steps each time. With the debounced keys of Fig 6.1 there is one sharp change-over from a high to a low output when the key is pressed. The NAND gates in this ic have Schmitt trigger inputs. They are normally held low by the resistors, and C1 (or C2) is charged. When the key is pressed, the input voltage rises steadily to high as C1 discharges. When the voltage reaches about 3.5 V the gate output goes low. The Schmitt input is such that, having reached this upper threshold of 3.5 V, the gate will not change back again unless the voltage falls as far as the lower threshold, which is about 2.7 V. If contact is broken briefly as the result of contact bounce, the charge on C1 prevents the voltage falling as far as 2.7 V. Thus having changed state *once* the gate does not change state again. The reverse action applies when the key is released; output rises sharply to high.

The C ('cancel') key, which is connected to the RESET inputs of the counters, does not need to be debounced, as it does not matter how many high pulses reach this input when the counter is being reset.

When key P is pressed, counter 2 is incremented and its outputs A to D go through a binary cycle from all low (0000, equivalent to zero) to 1001 (equivalent to 9). These outputs are decoded by IC3, which then generates the outputs required to switch on the segments of a 7-segment display, to show the corresponding decimal digit.

The logic levels from counter 2 are also fed to a set of four exclusive-OR gates (see Fig 0.2). Here they are matched against the output from four data selectors (IC6 and IC7). The matching is to check that the correct number has been keyed in at each stage. The data selectors are controlled by the output from counter 1. To begin with this is reset, with outputs A and B both low. This causes the

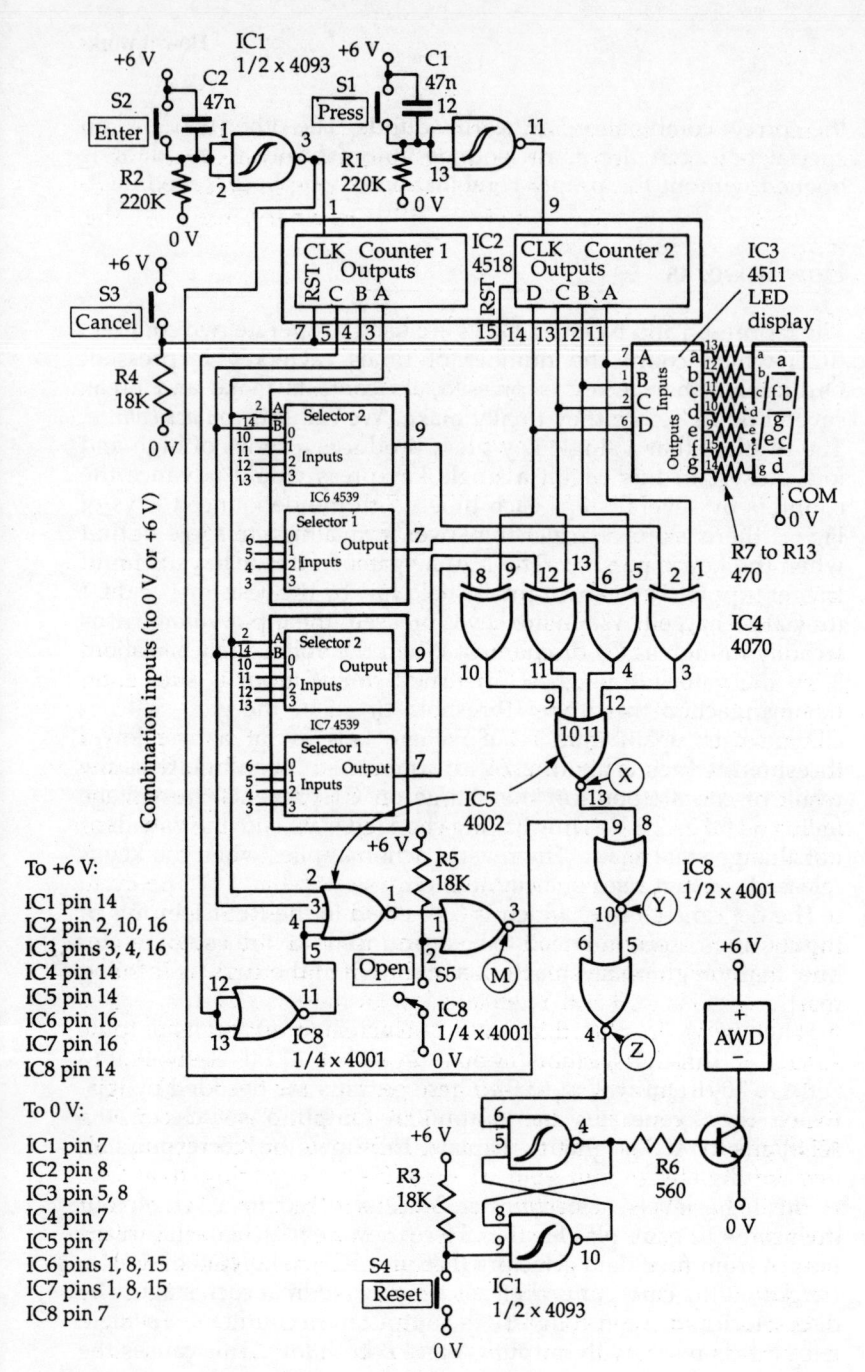

Figure 6.1 Circuit diagram of the combination door sentry

'0' input of each selector to appear at its output. The '0' inputs are each wired either to 0 V or +6 V, so that the outputs of the four selectors correspond to the first digit of the combination. For example, if the first digit is to be '5' (0101 in binary, see p. 8), the inputs are wired like this:

Binary digit	IC	Selector	Pin no.	Wire to
0	7	1	6	0 V
1	7	2	10	6 V
0	6	1	6	0 V
1	6	2	10	6 V

If the user has keyed in the correct number, each gate of IC4 receives either a pair of highs or a pair of lows, and the outputs of all 4 gates are low. These outputs are NORed together by IC5, and the output at pin 13 (referred to as X) is high. However, if the wrong number is keyed in, one or more of the gates of IC4 receives one low and one high input. Therefore, one or more of these gates has a high output and X is low.

X is fed to a NOR gate in IC8. When key E is not being pressed the gate receives a high input, so its output (Y) is low. It is low whatever the state of X. However, if key E is pressed to enter the number and X is low (wrong number), Y goes high. This is an alarm condition.

When key E is released, counter 1 steps on to state '1'. The input to the data selectors changes and the data is selected from the '1' inputs, to correspond with the second digit of the combination. The operator then keys in the second digit and enters it. If this is matching, all is well and Y remains low. If there is an error, Y goes high. Keying in the third and fourth digit follow the same sequence, matching against inputs '2' and '3' of the selectors.

After key E has been pressed and released for the fourth digit, the output of counter 1 changes to '100'. Output C is inverted by a NOR gate of IC8 so that the second NOR gate of IC5 receives all low inputs. Its output (pin 1) goes high, indicating that four numbers have been correctly entered. This is the stage at which the door must be opened. Opening the door closes S5 (we will explain how in a moment) and gives a low input to the next NOR gate. The output of this gate (M) is high if both inputs are low, that is, if the door is opened when counter 1 is not at stage 4. So, if the operator does not know how many digits the combination has, and tries to

Combination door sentry

open the door when other than four numbers have been entered, M goes high. This is another alarm condition.

M and Y are NORed together and the resulting output Z is low when either (or both) M and Y are high. Either alarm condition causes a low Z, which triggers the flip-flop (the cross-connected NAND gates of IC1). The output at pin 4, normally low, goes high, turns on TR1 and makes the audible warning device sound. It continues to sound until the flip-flop is reset by pressing S4. S4 is hidden in an inaccessible position (inside the protected room or cupboard) so that the alarm can not be silenced by unauthorised persons.

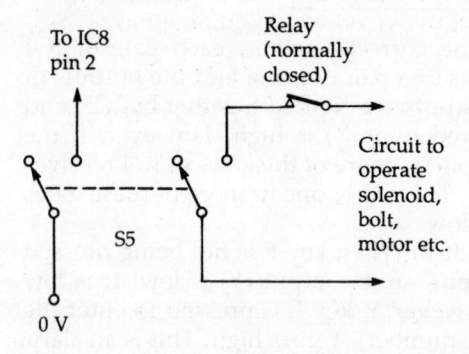

Figure 6.2 Alternative switching. The relay is optional

Switch S5 may be a magnetic switch or micro-switch on the door, arranged so that opening the door closes its contacts. Then any attempt to open the door, except when all four digits of the combination have been entered, makes the alarm sound. Alternatively, S5 may be a double-pole double-throw switch wired as in Fig 6.2. The second section of the switch operates a magnetically operated lock. It could instead turn on the power to any other electrically-powered device such as a motor, a lathe or a computer. In this way the sentry can be used to restrict the access to machinery or equipment to people who have been given the combination.

The optional relay in Fig 6.2 has its coil wired in place of the AWD. In a warning condition the contacts open, preventing the lock, machines or equipment from operating.

Construction

Begin with IC1 and the keys. You may need to drill extra holes for the key terminals if they are more than 1 mm in diameter, or not an exact multiple of 2.5 mm apart. Test outputs from IC2, then add IC2, IC3 and the display. The display can be soldered directly to the board, but it is preferable to mount it on two rows of socket-strip (e.g. Soldercon). Check that the display repeatedly advances from 0 to 9 as key P is pressed. Next add IC4 to IC7. Decide on the combination and connect the input terminals of IC6 and IC7 accordingly. The four digits of the combination are referred to as '0' to '3', and the corresponding input pins are marked 0 to 3 in Fig 6.3. The four binary digits are referred to as A to D. As an example, this is how we wire the combination '3529':

Digit reference	Value	Binary equivalent DCBA	Pins to 0 V IC7	IC6	Pins to 6 V IC7	IC6
0	3	0 0 1 1	6, 10	–	–	6, 10
1	5	0 1 0 1	5	5	11	11
2	2	0 0 1 0	4, 12	12	–	4
3	9	1 0 0 1	13	3	3	13

Test the circuit built so far by running through correct and incorrect keying sequences. X is normally high but goes low whenever a mistake is made.

Finally add IC8 and complete the connection to the flip-flop of IC1. Check that the circuit now operates as described above, and that the alarm is triggered by any mistake, or by closing S5 at any stage except immediately after the 4th digit has been entered.

The completed circuit may be housed in a plastic case fastened to the outside of the door. An aperture is cut in the panel of the case for the 3 keys and the display. The board is mounted just below the panel. S5 is mounted directly on the panel. The circuit requires a maximum of 30 mA, mainly for the display. It may be powered by a battery of four 'D' type cells, or a 6 V dc mains adaptor. To save current, leave the circuit with the display on '1'. This reduces power consumption to 15 mA.

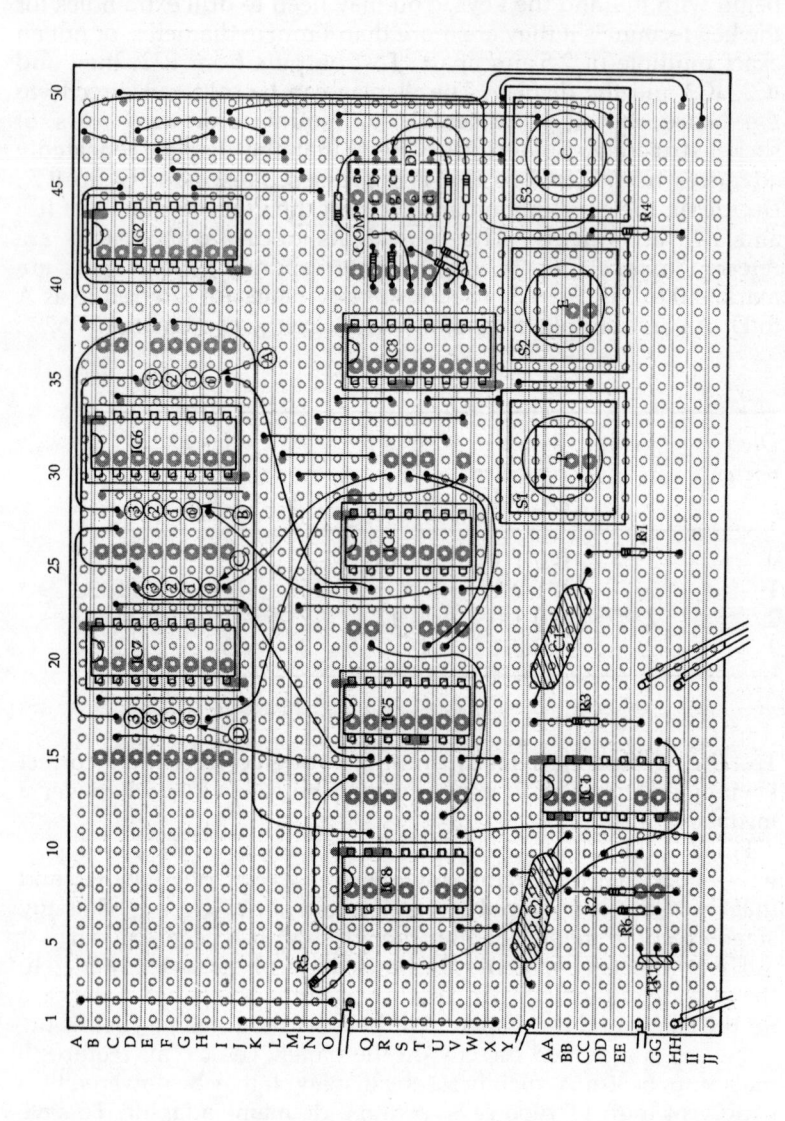

Combination door sentry

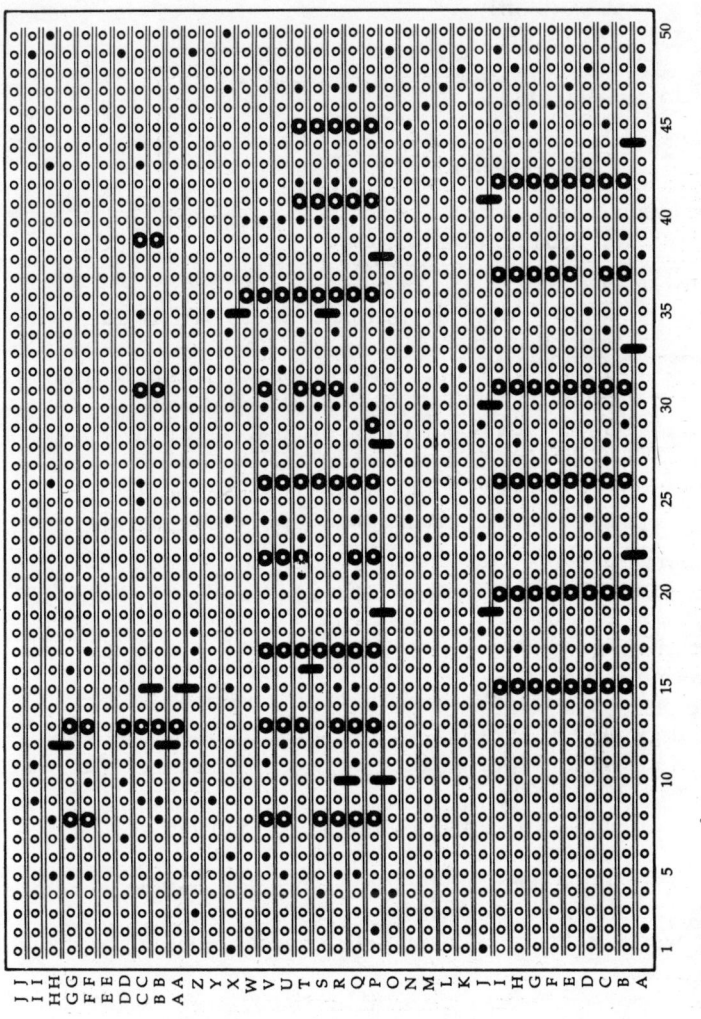

Figure 6.3 Stripboard layout of the combination door sentry

Components required

Resistors (see p.8)
R1, R2   220k (2 off)
R3-R5   18k (3 off)
R6   560
R7 to R13   470 (7 off)

Capacitors
C1, C2   47nF polyester

Semiconductors
TR1   ZTX300 npn transistor

Integrated circuits
IC1   4093B CMOS quadruple 2-input NAND gate with Schmitt inputs
IC2   4518B CMOS dual synchronous divide-by-10 counter
IC3   4511B CMOS 7-segment latch and driver
IC4   4070B CMOS quadruple exclusive-OR gate
IC5   4002B CMOS dual 4-input NOR gate
IC6,IC7   4539B CMOS dual 4-input data selector (2 off)
IC8   4001B CMOS quadruple 2-input NOR gate

Miscellaneous
LED 7-segment display, common cathode type (Fig 6.3 shows 0.3" display)
stripboard 127 mm x 95 mm (Vero 10347)
1 mm terminal pins (6 off)
suitable plastic enclosure
14-way d.i.l. sockets (4 off)
16-way d.i.l. sockets (4 off)
socket-strip (10 sockets)
S1-S3   keyboard switch, with key top (3 off)
S4   press-to-make push-button
S5   single-pole single-throw or double-pole double-throw toggle switch (see text)
audible warning device, 6 V (solid-state buzzer or siren), or relay

7 Digital die

This project is a good illustration of how to set about designing a digital logic circuit. The principle of the die is simple. An astable multivibrator generates a series of pulses at high frequency (about 5 kHz). The pulses are fed to a counter, which is wired to reset at the 6th count. The counter output cycles repeatedly through the sequence 0, 1, 2, 3, 4, 5, 0, 1, ... etc.

The output of the counter is binary. The function of the logic is to accept this 6-step binary output and to light an array of LEDs so as to produce the pattern of dots found on the faces of a conventional 6-sided die. Fig 7.1 shows the patterns required. When the counter is running, the LEDs are turned on and off so rapidly that it is not possible to see what score is showing. The circuit has a STOP

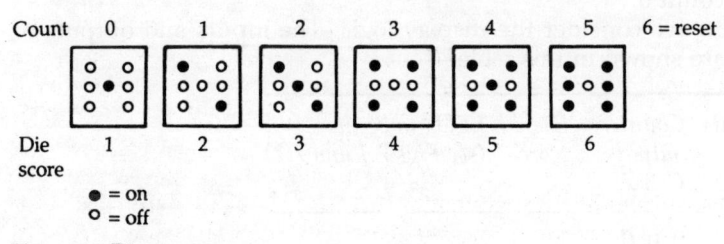

Count   0   1   2   3   4   5   6 = reset

Die score 1   2   3   4   5   6

● = on
○ = off

Figure 7.1 Die sequence

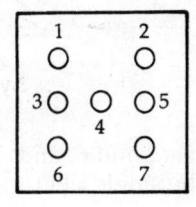

```
  1       2
  O       O
3O   O   O5
     4
  O       O
  6       7
```

Figure 7.2 Numbering of LED'S of die display

59

Digital die

button which halts the counter when pressed. The display then shows one of the scores 1 to 6. The operator has no way of knowing what the score will be when the button is pressed so, in effect, the score is obtained just as randomly as when rolling a conventional die. This project has many uses as a novel die for all kinds of die-based games and has the advantage that it will never roll off the table and disappear under the sofa!

Designing the logic

Since we are dealing only with numbers up to 6, we need be concerned only with the first three stages of the counter. We refer to these as A, B and C (Fig 7.3). First of all we have to make the counter reset to zero immediately it reaches a count of 6. In binary, the value 6 is 110, in which output A is low and outputs B and C are high. The table on p. 8, shows that this is the *first* stage at which both B *and* C are high. Therefore we use logic AND to detect this stage. The AND gate at the top right of Fig 7.3 takes the B and C outputs from the counter and ANDs them. The 'BC' beside the output of this gate is a short-hand way of writing 'B AND C' (it is in fact an example of the notation used in Boolean algebra). When B and C go high, the output of the gate goes high (see Fig 0.2). The high output resets the counter. This all happens in a few milliseconds so that the counter apparently goes direct from count 5 to count 0.

Now we consider the display logic. The inputs and outputs of this are shown in this table:

Count	Counter outputs C B A	Die score	LEDs to be lit (see Figs 7.1 and 7.2)
0	0 0 0	1	4
1	0 0 1	2	1 7
2	0 1 0	3	1 4 7
3	0 1 1	4	1 2 6 7
4	1 0 0	5	1 2 4 6 7
5	1 0 1	6	1 2 3 5 6 7

LEDs 1 and 7 always come on together, as do LEDs 2 and 6, and LEDs 3 and 5. With these three pairs of LEDs and the single LED 4

Designing the logic

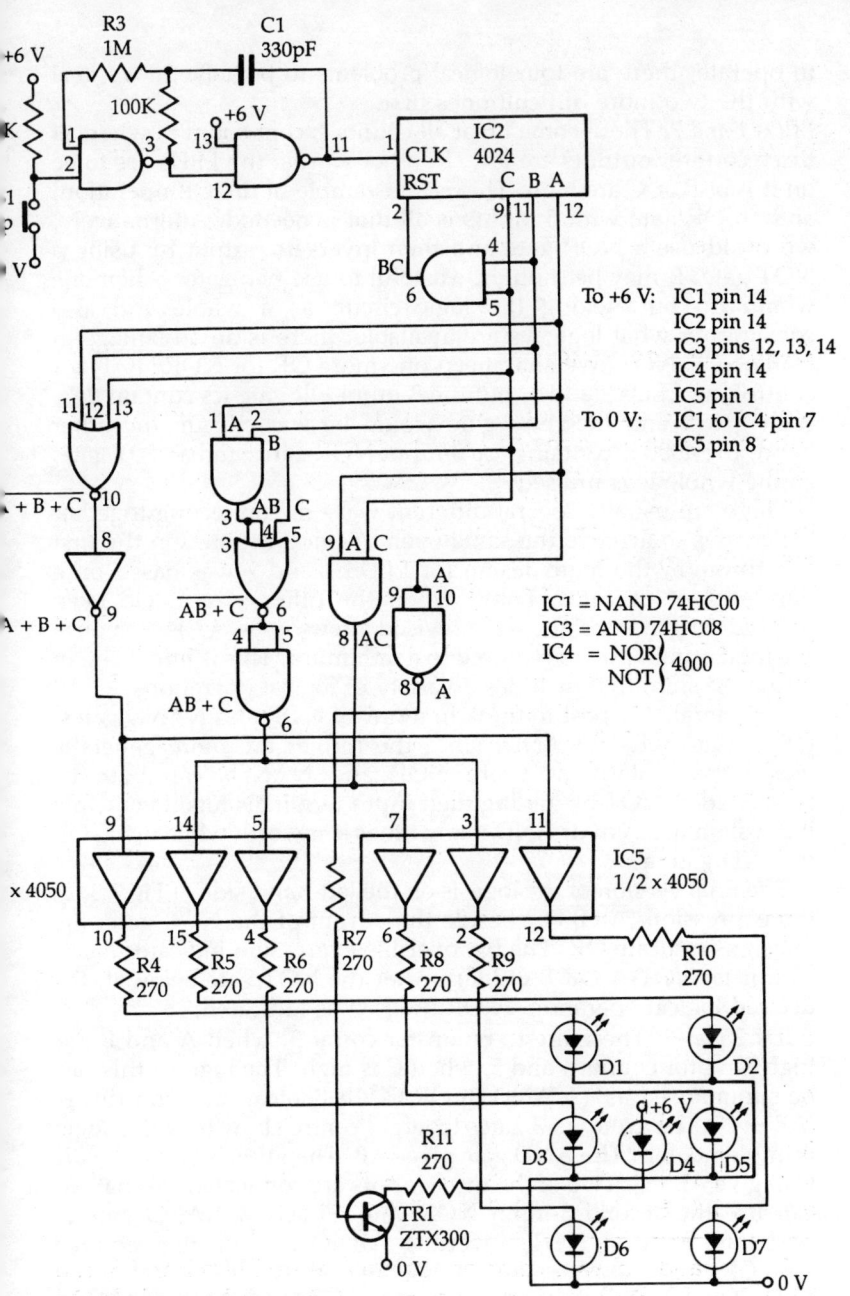

Figure 7.3 Circuit of the digital die

Digital die

to operate, there are four logical problems to be solved. We deal with the two more difficult ones first:

LEDs 1 and 7: These come on at all counts except count 0, when all three counter outputs are low. In other words, the LEDs are to be on if A *or* B *or* C are high. This is an example of the OR operation, and an OR gate with 3 inputs is all that is needed. Alternatively, we could use a NOR gate and then invert its output by using a NOT gate. It may be thought wasteful to use two gates when one would do but, taking the logic circuit as a whole and also considering what logic ics are available, there is an advantage in using NOR/NOT. We shall need one more OR (or NOR/NOT) for controlling LEDs 2 and 5, but the 3-input OR gate ics contain *three* such gates, and the third gate would be wasted. On the other hand, the 4000 ic contains *two* 3-input NOR gates and a NOT gate, so the whole ic is utilised.

There are usually several different ways of connecting together logic gates to achieve the same overall logical action. On the first run through, the logic design for LEDS 1 and 7 was based on a different set of gates. Later, when the other stages had been worked out, the designs were revised a few times, so as to reduce the total number of ics required to a minimum. This is important in a small system that includes a variety of logical operations.

In general, it is best to think in terms of NOR and NAND gates, rather than OR and AND, since the former are more generally useful (particularly as spare NOR and NANDs can both be converted to NOT by joining their input terminals together). Only if a system uses OR or AND many times is it worth while using OR or AND gate ics.

The final version of the logic is on the left-hand side of Fig 7.3. In the expression '$\overline{A+B+C}$' beside the output of the NOR gate, the '+' signs indicate OR. The bar over the expression indicates NOT, so this is 'NOT(A OR B OR C)'. After the NOT gate below it, the desired logical operation 'A OR B OR C' is indicated.

LEDs 2 and 6: These are to be on for count 3, when A and B are high, and for counts 4 and 5, when C is high. The logic of this can be summarised as '(A AND B) OR C'. In Boolean notation this is 'AB + C'. The group of 3 gates to left of centre show how this logic is implemented. The AND gate gives AB. Then this is NORed with C to give $\overline{AB+C}$. Two of the three inputs are connected, so that we can use the second 3-input NOR gate. Finally a NAND gate is wired to perform NOT, inverting $\overline{AB+C}$ to give the required AB+C.

LEDs 3 and 5: these come on only at count 5, when A *and* C are

Designing the logic

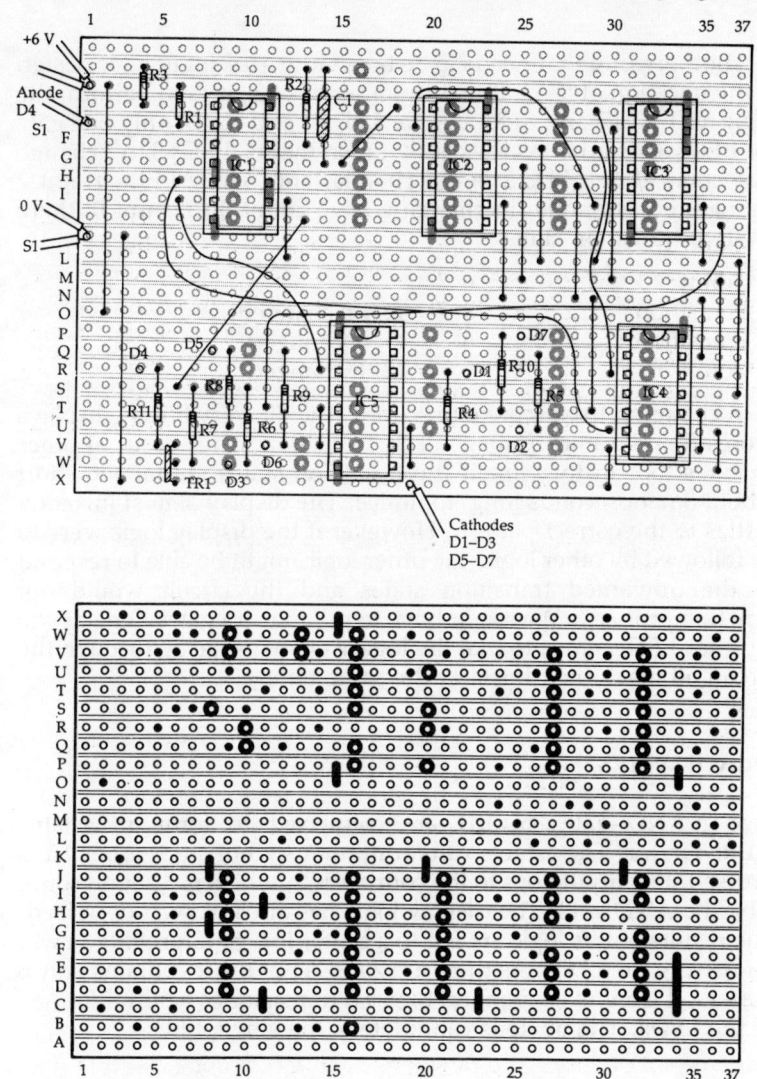

Figure 7.4 Stripboard layout of digital die

high, so the logic is 'A AND C' or AC. The gate in the centre of Fig 7.3 shows the solution.

LED 4: This lights at the even counts 0, 2 and 4. All we need to do is to invert A, to give NOT A ($\overline{A}$). There is a spare NAND gate that is used for this purpose.

Digital die

Another point that needs to be taken into account in logic design is the question of *races*. It requires a series of 3 gates to perform the logic for LEDs 2 and 6, 2 gates for LEDs 1 and 7, but only one gate for LEDs 3, 4 and 5. As the counter increments, LEDs 3 to 5 change state first, then LEDs 2 and 6 change, and finally LEDs 1 and 7 change. For example, in changing from count 5 to count 0, there are two transition states:

Count 5	1	2	3		5	6	7
Transition	1	2		4		6	7	(3,4,5 changed)
Transition		2		4		6		(1, 7 changed)
Count 0				4				(2, 6 changed)

In changing from score 6 to score 1, the die very briefly displays scores 5 and a mirror-image of 3. Transition states occur at other changes too. In this circuit the transition states are far too short (about 8 nanoseconds long) to notice. The display almost instantly settles to the correct pattern. However if the display logic were to be followed by other logic, the other logic might be able to respond to the unwanted transition states and the circuit would not function properly. We might have to put in extra gates to delay the logic for LEDs 3 to 5, to that the logical changes racing through the circuit all reach the 'finish' at the same time.

How it works

Two NAND gates of IC1 are connected to form an astable multivibrator. The astable runs continuously but stops when S1 is pressed, Its output goes to the counter IC2, which is reset on count 6 by the AND gate. The display logic has already been described. The results of this logic are fed to a buffer ic IC5, to provide power for driving the LEDs. As there are 7 LEDs but the ic contains only 6 buffers, it is more economical to use a transistor as a buffer for one of the LEDs. The obvious candidate is the single LED 4.

Construction

Build the multivibrator and connect the counter to it. Monitor the outputs of the multivibrator and counter using an oscilloscope or, more simply, by using the devices of Project 1. Complete the logic circuits and buffers. The LEDs are set out in an array (Fig 7.2) on the lid of the plastic case. The push-button and the main power switch are mounted beside the array. The circuit requires only 25 mA, so it can be driven by four AA cells, or a 6 V battery (PP1).

When completed, the circuit may be tested with a 2.2 µF capacitor connected *in parallel with* C1. This slows down the multivibrator so that the sequence of display patterns can be checked. If is does not run through all the patterns shown in Fig 7.1, check that the LEDs are soldered in the right way round. If one set of LEDs seems to be operating incorrectly, check the wiring to the gates concerned.

Components required

Resistors (see p.8)
R1   18k
R2   100k
R3   1M
R4-R11   270 (8 off)

Capacitors
C1   330pF polystyrene
C2   2.2uF (for testing)

Semiconductors
D1-D7   TIL209 or similar LEDs
TR1   ZTX300 npn transistor

Integrated circuits
IC1   74HC00 CMOS quadruple 2-input NAND gate
IC2   4024B CMOS 7-stage binary ripple counter
IC3   74HC08 CMOS quadruple 2-input AND gate
IC4   4000B CMOS dual 3-input NOR gate plus inverter
IC5   4050B CMOS hex non-inverting buffer

Miscellaneous
S1   push-to-make push-button
S2   single-pole single throw toggle switch
stripboard 95 mm x 63 mm (Vero 10346)
1 mm terminal pins (11 off)
suitable plastic enclosure
14-way d.i.l. sockets (5 off)
16-way d.i.l. socket
battery box/battery connector (optional)

8 A Christmas decoration

This project, with its two optional extensions, will add a novel touch to your seasonal decor. The circuit can be applied in many ways, and and there is plenty of scope for you to show your creativity in producing a design that is truly original and unique.

The basic circuit makes a set of 5 LEDs light up according to the binary sequence, repeating every quarter-minute. The LEDs are mounted on a festive decoration of your choice. You can buy a decoration ready-made or design and make your own. It can be hung on the Christmas tree or displayed elsewhere in the room.

The first extension adds 4 more LEDs to the display. The LEDs light up one at a time until all 4 are lit. They stay lit for a while, then go out one at a time. This action repeats once every half-minute. The second extension adds sound. It employs a special music-generating ic, intended for use in musical greeting cards. There are 4 possible versions of this. The version chosen for this project plays a 64-note medley of 'Jingle Bells', 'Santa Claus is coming to town' and 'We wish you a Merry Christmas'. The music is heard every 4 minutes. The circuit allows a degree of flexibility in timing, so the tune can be played more or less frequently - even as infrequently as once an hour. There are also many possible variations in the timing of the lighting changes, the number of LEDs and their colour.

How it works

The astable multivibrator, or clock (IC1, Fig 8.1), runs at about 1 Hz. It drives D1 and a binary counter (IC2). This lights four more LEDs D2 to D5. The 5 LEDs light according to a repeating binary sequence from 00000 to 11111. Further LEDs can be connected to the other outputs of IC2 if required.

How it works

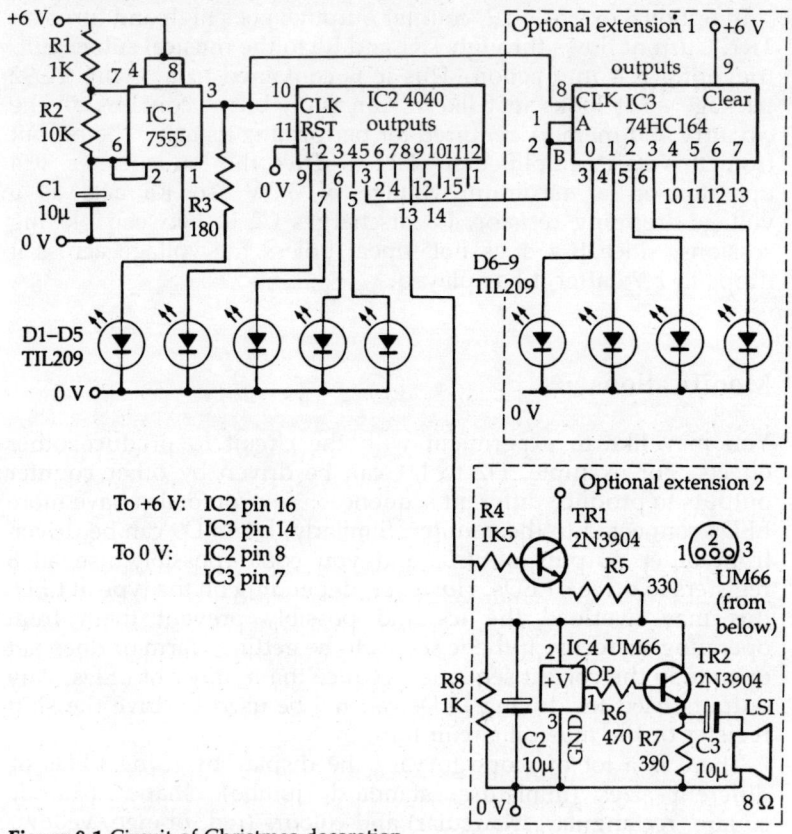

Figure 8.1 Circuit of Christmas decoration

In optional extension 1, the output from stage 5 of the counter is taken to the serial input of a shift register (IC3). This contains a chain of 8 linked registers. Each time the clock goes high the data in each register is shifted to the next register along the chain. The first register acquires the data presented to it from stage 5 of the counter. Data in the 8th register is lost. When any of the first 4 registers contain a 'high' the corresponding LED lights. The effect is that, when output 5 of IC2 is high, the registers 1 to 4 become high in turn as data is shifted at each clock count. One by one, the LEDs D6 to D9 come on. They stay on until output 5 of IC2 goes low. Then they go out, one by one.

A Christmas decoration

In optional extension 2, counter output 8 goes high and turns on TR1. Current flows through TR1 and R5 to the musical sub-circuit, triggering IC4 into action. This ic is contained in a 3-wire TO92 package so it looks just like a transistor, but it contains all the circuitry and memory required for performing its task. The output from this IC is amplified by TR2 to drive the loudspeaker. IC4 operates on a maximum voltage of 3.3 V, so R5 acts as a voltage-dropping resistor. R8 discharges C2 in between playing sessions, since IC4 does not repeat unless the voltage across it drops to zero after it has played.

Modifications

You may like to experiment with the circuit to produce other effects. For example, D2 to D4 can be driven by other counter outputs to produce different sequences. You could also have more LEDs connected to the counter. Similarly, D6 to D9 can be driven from other outputs of IC3, and you could possibly use all 8 registers to drive 8 LEDs. However, depending on the type of LED, this may overload the ics and possibly prevent them from operating correctly. If the ic seems to be getting warm or does not operate in the correct sequence, reduce the number of LEDs. Any outputs used for driving LEDs can not be used to drive the shift register or the musical circuit too.

There is a lot of scope to vary the display by using LEDs of different sizes (miniature, standard, jumbo), shapes (round, square, rectangular, triangular) and colours (red, orange, yellow, green). Another interesting variation is to try a tri-colour LED. This has a red LED and a green LED in the same package, with 3 terminal wires so that they can be switched on independently. If both LEDs are on together, it appears yellow. A tri-colour LED connected to two binary outputs goes through the sequence off-red-green-yellow-off.

Construction

Fig 8.2 shows the stripboard layout. Build the clock and temporarily connect D1. The LEDs go on and off once a second. Add the counter and temporarily connect the LEDs. These light in a binary sequence (see p. 8). If you are adding extension 1, connect

Construction

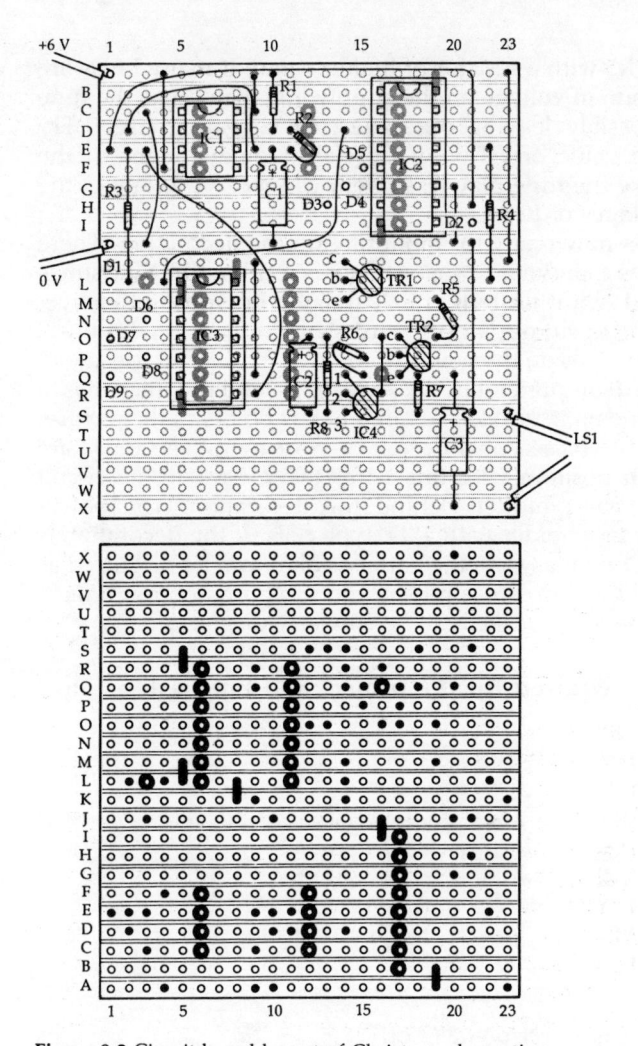

Figure 8.2 Circuit-board layout of Christmas decoration

IC3 and temporarily connect D6 to D9. They light up one at a time, stay on for about 14 seconds, and then go out one at a time.

If you have built extension 2, the tune plays once each time power is applied to it. Run the circuit and wait until output 8 (IC2 pin 13) goes high. The tune begins and plays through once. While it is playing, check the voltage at pin 2 of IC4. If this is more than

69

3.3 V, replace R5 with a resistor of higher value. If many LEDs are used, fluctuation in voltage on the supply line may cause the tune to hiccup, possibly leaving out notes, or repeating itself. Try increasing the value of C2 or, if this fails, remove some of the LEDs. To make the tune play less often, connect R4 to one of the later output stages of IC2.

The circuit is now ready for mounting on the decoration. There are many types that would prove suitable. A miniature Christmas tree festooned with flashing LEDs would be most attractive. Another idea is to cut out a cardboard star, decorate it with tinsel and mount the LEDs on the points of the star. You could use an old Christmas card or preferably an Advent calendar which has a Christmas scene on it, and locate LEDs at key points in the picture. Or draw a scene or seasonal design of your own. The LEDs are easily glued in position with fine wires taken back to the circuit board. This is best hidden behind the decoration. The power supply is a battery holder with 4 'D' type cells. If the decoration is hung on the Christmas tree, the battery holder can be disguised as a gift, wrapped in Christmas paper, and hung on another branch of the tree. The loudspeaker can be similarly disguised.

Components required for the project and both extensions

Resistors (see p.8)
R1	1k
R2	10k
R3	180
R4	1k5
R5	330
R6	470
R7	390
R8	1k

Capacitors
C1-C3   10 µF electrolytic (3 off)

Semiconductors
D1-D9   TIL209 or similar LEDs (9 off)
TR1,TR2   2N3904 npn transistor (2 off)

Integrated circuits
IC1   7555 CMOS timer
IC2   4040B CMOS 12-stage ripple counter
IC3   74HC164 CMOS 8-bit shift register
IC4   UM66 melody generator, type 1

Components required for the project and both extensions

Miscellaneous
LS1　　　　　loudspeaker, 8 ohms
stripboard 58 mm x 63 mm (cut from Vero 10346)
1 mm terminal pins (13 off)
8-way d.i.l. socket
14-way d.i.l. socket
16-way d.i.l. socket
6 V battery holder for 4 'D' type cells
materials for making decoration

9 Weekly reminder

This project is a reminder of anything that is due to happen on a weekly basis. It reminds you to return your library book, switch on your favourite TV programme, put out the dust-bin, leave the money for the milk vendor, clean the car, or phone your best friend. On the same day each week, it automatically flashes its LED to prompt you that 'something' has to be done. The LED flashes all day, if necessary, until you have done whatever you have to do, and have pressed the reset button. Rather than rely on a long-period timing circuit, with its problems of keeping to the exact time over periods of months, we have based the circuit on a much more reliable time-keeper – the Sun. This circuit literally counts the days. It is specially designed to have low power requirements – only 0.4 mA for most of the time – so it runs for months on a battery of four type AA cells.

How it works

The rising of the Sun is detected by a phototransistor (TR1, Fig 9.1). When the Sun rises, the increasing light level causes an increasing current to flow through the transistor. This current, also flows through R1 and VR1 and produces an increasing potential difference across them. The voltage at point A gradually falls. Eventually it falls to a level at which it is accepted as a low by Gate 1. The day has begun. Since the other input to this gate is high, its output changes to high. This is inverted by gate 2, the output of which goes low and makes the counter (IC2) step on one count. The day has been counted.

Gate 1 has Schmitt trigger inputs (p.51) so, once day has begun, a reduction in light due to increasing cloud, for example, is not

How it works

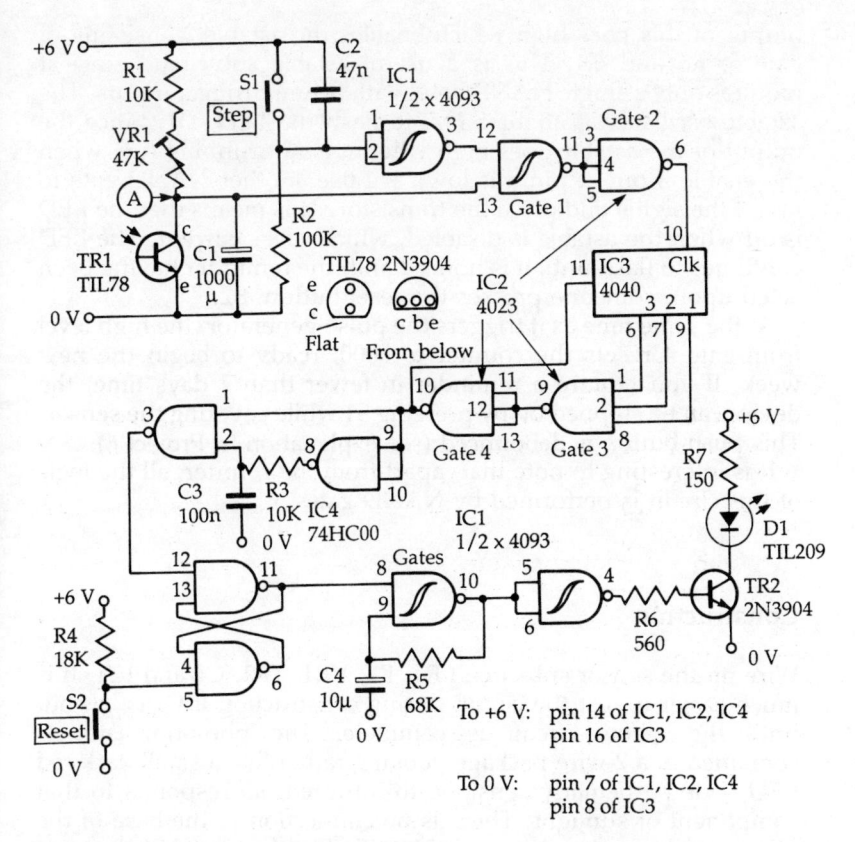

Figure 9.1 Circuit of the weekly reminder

enough to make the gate change back to its night-time state. Just in case the shadow of a passer-by should fall on the sensor during the day, the capacitor C1 damps out any sudden changes of voltage at A. Thus the circuit makes no further changes in response to light intensity until darkness falls at the end of the day. Similarly, the light from passing cars, lightning flashes and other transient nocturnal phenomena will not make the device think that the next day has arrived.

First thing every morning, the counter is incremented until it has counted seven days. On the morning of the seventh day, all inputs to gate 3 are high. Its output goes low, is inverted by gate 4 and triggers the pulse-generator (the next two gates). This produces a very short low pulse which in turn triggers the flip-flop. The

73

output of this goes high which enables the astable, consisting of gate 5, R5 and C4. This is a useful astable sub-circuit since it requires only a single NAND gate, with Schmitt trigger inputs. The astable oscillates at about 2 Hz, to flash the LED, D1. Since the output of the astable goes high when it is not running (i.e. when the enable input at pin 8 is low), we use another NAND gate to invert the signal and drive the transistor. This means that the LED is off when the astable is disabled, which saves current. The LED continues to flash until it is noticed and, the reminder having been acted upon, someone presses the reset button, S2.

At the same time as it triggers the pulse-generator, the high level from gate 4 resets the counter to 0000, ready to begin the next week. If you require a reminder in fewer than 7 days time, the device can be stepped on by pressing S1 while covering the sensor. This push-button is debounced (see explanation in Project 6).

It is interesting to note that, apart from the counter, all the logic of this circuit is performed by NAND gates.

Construction

Wire up the sensor sub-circuit R1, R2, VR1, TR1, C2 and IC1. It is much easier to test the circuit during construction if C1 is left out until the whole circuit is complete. The phototransistor is contained in a 2-wire package, looking rather like a small dark red LED. It is particularly sensitive to infra-red, so responds to that component of sunlight. There is no connection to the base of the phototransistor; the collector is identified by the 'flat'. IC1 does not work reliably if the other two gates are unconnected, so wire up these now, though they will be tested later. Test the sensor sub-circuit by shading TR1, measuring the voltage at point A (the collector wire of TR1) at the same time. It is low (about 0.1 V) in bright light but rises almost to 6 V in complete darkness. Also measure the output from pin 11 of IC1, possibly using the logic level probe (Project 1). This is high in light, low in dark; adjust VR1 so that the changeover occurs at a suitable light level. Test the operation of S1 by covering the sensor (pin 11 goes low), then pressing S1. Pin 11 goes high.

Next connect IC2 and IC3. Test these by monitoring the output of IC2 pin 10 while covering and uncovering the sensor repeatedly. The output goes high on every 7th *un*covering of the sensor. Complete the remainder of the circuit, including a power switch in the wire joining the positive terminal of the battery to the circuit. The LED usually flashes when power is switched on. Pressing S2 stops the flashing. Cover the sensor and press S1 repeatedly until the LED starts to flash. This brings the counter to step 7. Press S2 again to stop the flashing. Uncover and cover the sensor repeatedly; the LED starts flashing on the 7th uncovering.

The circuit is enclosed in a plastic case which could be a small one with an integral battery compartment for four AA type cells. On the rear of the case (assuming that the device is to be kept on a windowsill where it can receive daylight) bore a hole about 1 cm in diameter to allow light to reach the sensor. Another hole is required for the LED. Fig 9.2 shows the LED mounted on the circuit-board but it may be more convenient to mount it on the wall of the case and run leads to it from terminal pins at J14 and K14 on the board. It may possibly be more convenient if the sensor is mounted outside the case, say on a window-frame, so that it is not shaded when the curtains are drawn together. It can be glued there, or held in a lump of Blu-tack. Run leads from the sensor to terminal pins at C3 and D4 on the board.

Holes are also cut in the case for mounting the power switch and the two push-buttons. It is useful if a hole about 4 mm in diameter is bored in the case just above the centre of VR1. This allows a small screwdriver to be inserted to adjust VR1 to set the light level at which the device responds.

Setting up

Switch on the power; the LED usually flashes; if so, press S2 to stop it. Cover the sensor and press S1 several times until the LED flashes again. The counter is now at a count of 0 and is ready to start its weekly sequence. Press S2 to stop the flashing. If you uncover the sensor immediately, the device registers the arrival of the next day and it flashes the LED after only 6 days (though weekly after that). To make it flash weekly on the same day of the week as that on which it is set up, keep the sensor dark until night. Then expose it in darkness, ready to respond to sunrise.

To set it to operate in fewer than 7 days time, proceed as before, keeping the sensor covered, until the counter is at step 0. Uncover the sensor *in the light* – this sets it to flash in 6 days time, as explained above. Pressing the STEP button (S1) further reduces the period by a day for each press.

Weekly reminder

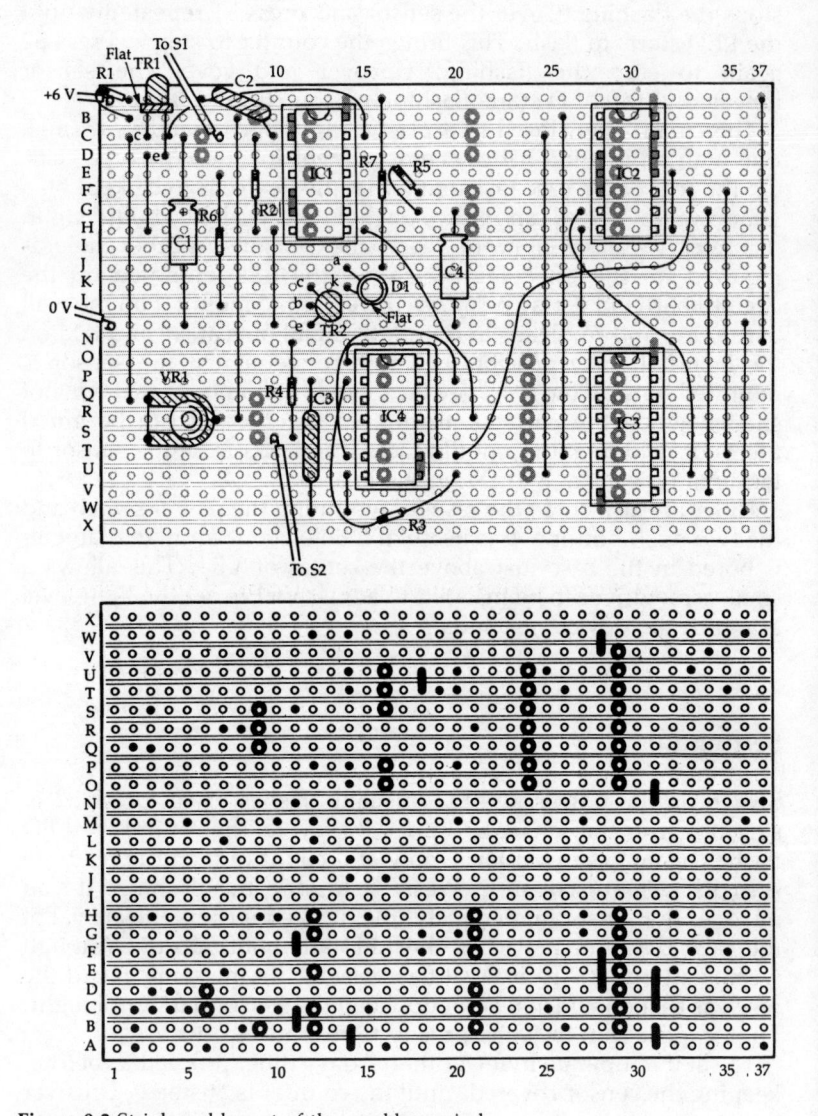

Figure 9.2 Stripboard layout of the weekly reminder

Components required

Resistors (see p.8)
R1,R3　　　10k (2 off)
R2　　　　　100k
R4　　　　　18k
R5　　　　　68k
R6　　　　　560
R7　　　　　150
VR1　　　　47k miniature horizontal preset potentiometer

Capacitors
C1　　　　　1000μ electrolytic
C2　　　　　47n polyester
C3　　　　　100n polyester
C4　　　　　10μ electrolytic

Semiconductors
D1　　　　　TIL209 or similar LED
TR1　　　　TIL78 phototransistor
TR2　　　　2N3904 npn transistor

Integrated circuits
IC1　　　　　4093B CMOS quadruple 2-input NAND gate with Schmitt trigger inputs
IC2　　　　　4023B CMOS triple 3-input NAND gate
IC3　　　　　4040B CMOS 12-stage binary ripple counter
IC4　　　　　74HC00 CMOS quadruple 2-input NAND gate

Miscellaneous
S1, S2　　　push-to-make push-button
S3　　　　　single-pole single throw toggle or slide switch
stripboard 95 mm x 63 mm (Vero 10346)
1 mm terminal pins (4 off)
suitable plastic enclosure
14-way d.i.l. sockets (3 off)
16-way d.i.l. socket
battery holder for four AA cells (if case does not have integral compartment)
battery clip

10 Remote-control switch

This device consists of two relay switches which you can control by clapping your hands, by flicking your fingers, or by whistling. To begin with, both switches are off. As each clap is detected by the device, the two switches go though the following sequence:

Clap	Switch 2	Switch 1
Start	Off	Off
1	Off	On
2	On	Off
3	On	On
4	Off	Off

This sequence corresponds to the binary numbers 00 to 11 (see p.8). Depending on the exact type of relay used, other switching actions are easily obtainable with very little extra wiring. The device has many applications in robotics and in model control. For example, it can be used to start and stop a model locomotive and control a set of points. Since the switches are relays, it can be used for controlling mains-powered equipment. A useful example is a porch light that comes on when you clap. This is much easier than fumbling in darkness to find the switch. It also has the benefit that a noise made by anyone attempting to force the door at night turns on the porch light, and exposes the culprit to view. Wired to a lighting fitting with two bulbs, it can be used to switch on either of the bulbs singly or both together. Or it can control one lamp and an electric fan. However, it must be emphasised that wiring up the mains section of such a circuit, though simple to do, is not a task for the absolute beginner. If you want to try your hand at mains applications, ask the advice of an experienced friend.

How it works

Sound is detected by a microphone (XTAL1, Fig 10.1) and causes an alternating voltage to appear at the junction of R3 and R4. This is amplified by the operational amplifier (IC1), the output of which is rectified by the diode D1. As the output voltage swings positive, current passes through D1 increasing the charge on C2. As the voltage swings negative, D1 prevents current from passing back to the amplifier. Thus noise causes a build-up of charge on C2, and an increasing voltage at the input to the second amplifier IC2.

The charge on C2 leaks away steadily through R8. With normal noise levels in a room the voltage does not rise appreciably. But with relatively loud noise, with a lot of high-frequency content, the voltage rises steeply. IC2 compares the voltage on C2 with a steady voltage at the wiper of VR1. This steady voltage is set so that it is just above the normal voltage on C2. Under such circumstances, the output from IC2 is high. When a clap is detected, the voltage on C2 rises above that at the wiper of VR1, causing the output of IC2 to swing sharply to 0 V. This triggers IC3, which is wired as a monostable multivibrator. Its output, normally low, goes high for a period of about 0.2s. This gives time for the effect of the original sound and of any subsequent echoes from the walls of the room to die away. We have a single response for a single clap.

The output from IC3 goes to a counter, built up from two D-type flip-flops, IC4. The operation of a D-type flip-flop is as follows. When the clock input of the first flip-flop changes from low to high (i.e. when a clap is detected) the data that is present at the D input appears at the Q output. The inverse of the data appears at the Q̄ output. In this circuit the Q̄ output is wired to the D input. The effect of this is that the flip-flop changes to the opposite state each time it is clocked, i.e. at each clap. A second flip-flop is clocked by the Q̄ output of the first flip-flop, so it changes state at half the rate of the first flip-flop.

The Q outputs of the flip-flops are fed to two transistors, which have the relay coils in their collector circuits. When a flip-flop output goes high, the transistor is switched on and the coil is energised. The diodes D2 and D3 are there to protect the transistors from the damaging effects of back e.m.f. when the current through the relay coil is turned off.

Remote-control switch

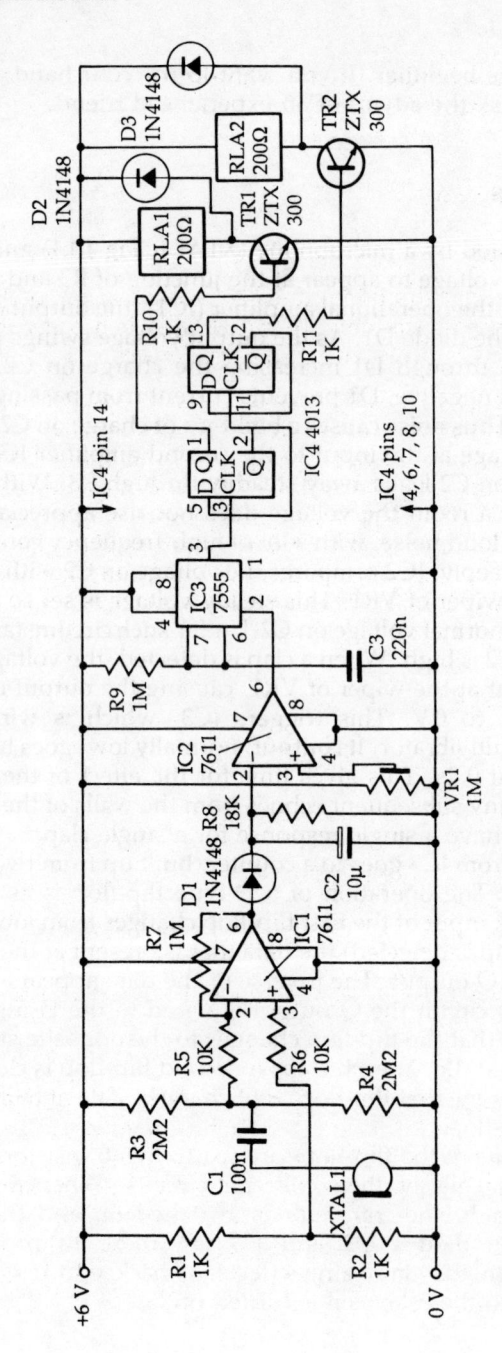

Figure 10.1 Circuit of the remote-controlled switch

The relays are used to switch one or more devices on or off, depending on the application.

Choosing the relays

Two points apply to any of the relays to be used with this project:

(1) The relay coil should be rated to operate at 6 V or 12 V dc.
(2) The relay contacts must be rated to switch the voltage and current required to drive the controlled devices.

If you require only a simple 'on/off' action, you need only one relay, with single change-over contacts (Fig 10.2). R11, TR2 and D3 are omitted from the circuit. In Fig 10.2, the symbol shows the position of the moving contact when the coil is *not* energised. The moving contact is held by a spring that presses it against one of the fixed contacts. This contact is unconnected, so the relay switch is off. When the coil is energised, the moving contact is pulled across to the other fixed contact. This completes the circuit and the lamp is turned on.

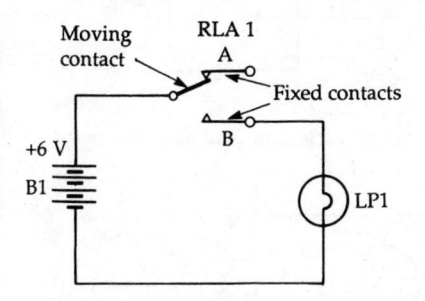

Figure 10.2 A relay with changeover contacts being used to switch a lamp on or off (relay coil not shown)

Many types of relay have two independent pairs of change-over contacts. These can be used for switching two devices on or off simultaneously. Or, as in Fig 10.3, we may use such a relay as a reversing switch for a dc motor. In Fig 10.4 we have two relays, one as an on/off switch (as in Fig 10.3) and the other as a reversing switch. The sequence of operation is 'off, forward, off, reverse'. Two claps in rapid succession produce a quick change from forward to reverse without apparently stopping the motor.

Remote-control switch

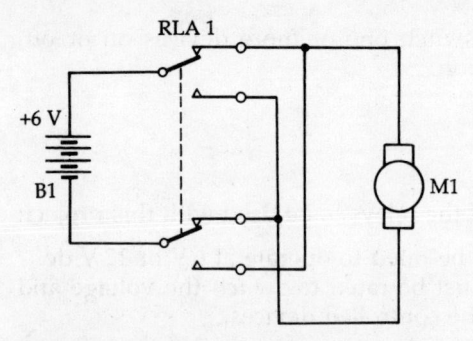

Figure 10.3 A relay with double changeover being used as a reversing switch for a d.c motor

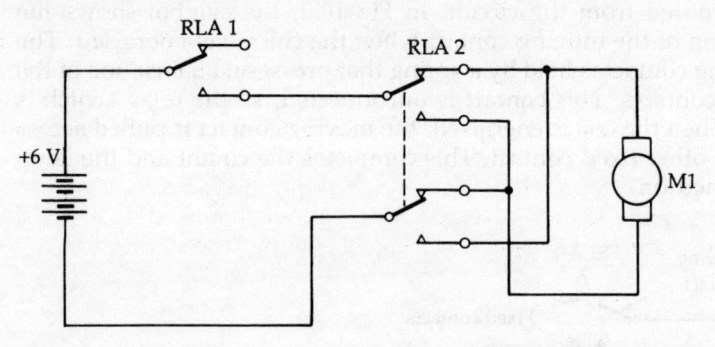

Figure 10.4 Using two relays to obtain on/off/reverse control of a d.c motor

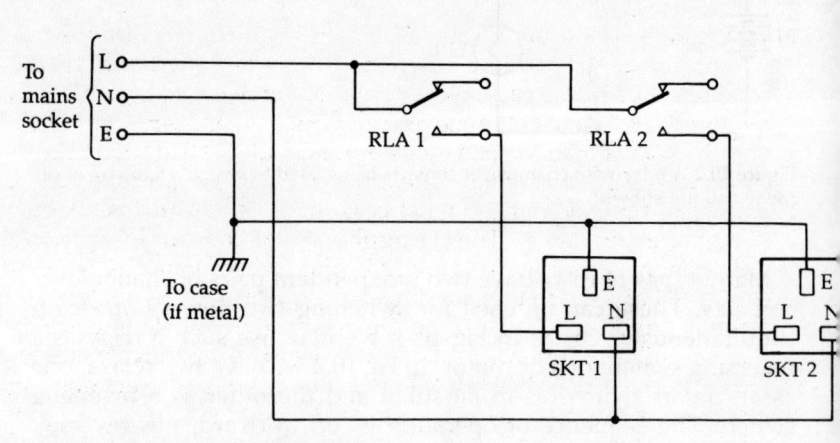

Figure 10.5 Connections for switching two mains-powered sockets (L = live; N = neutral; E = earth

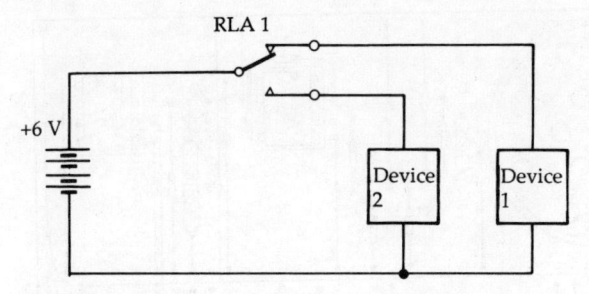

Figure 10.6 Using a single relay to switch on two devices, one at a time

Fig 10.5 shows one way of using a pair of relays with change-over contacts to control two devices plugged into two 3-pin mains sockets. You could of course use just one relay (RLA1) if you have only one device to control (e.g. a porch lamp). With the connections as in Fig 10.5, the switching sequence is 'both off, device 1 on, device 2 on, both on '. If the two devices are lighting bulbs, we have a simple method of dimming room lighting.

It is easy to use a single relay to switch one device on and another off at the same time. Fig 10.6 shows the wiring.

Construction

The project works on 6 V or 12 V. On 6 V it requires only 8 mA when quiescent but more when the relays are energised. These require about 27 mA each, if they have 200-ohm coils but, with coils of higher resistance, the current required is smaller. On a 12 V supply it takes 16 mA when quiescent, and about 60 mA per relay if these have 200-ohm coils. The project could be powered by a battery of four (or eight) type D cells, or a 300 mA mains-adaptor ('battery eliminator'). The microphone can be an inexpensive 'microphone insert'. In Fig 10.7 the microphone is shown connected directly to the board by two very short leads soldered in holes F11 and H11. The body of the microphone is held in place by a piece of Blu-tack between it and the board. Alternatively, solder terminal pins at F11 and H11 and run leads to a microphone mounted off-board. This is preferred, for example, if the microphone is to be mounted outdoors in a porch while the rest of the circuit is indoors.

Remote-control switch

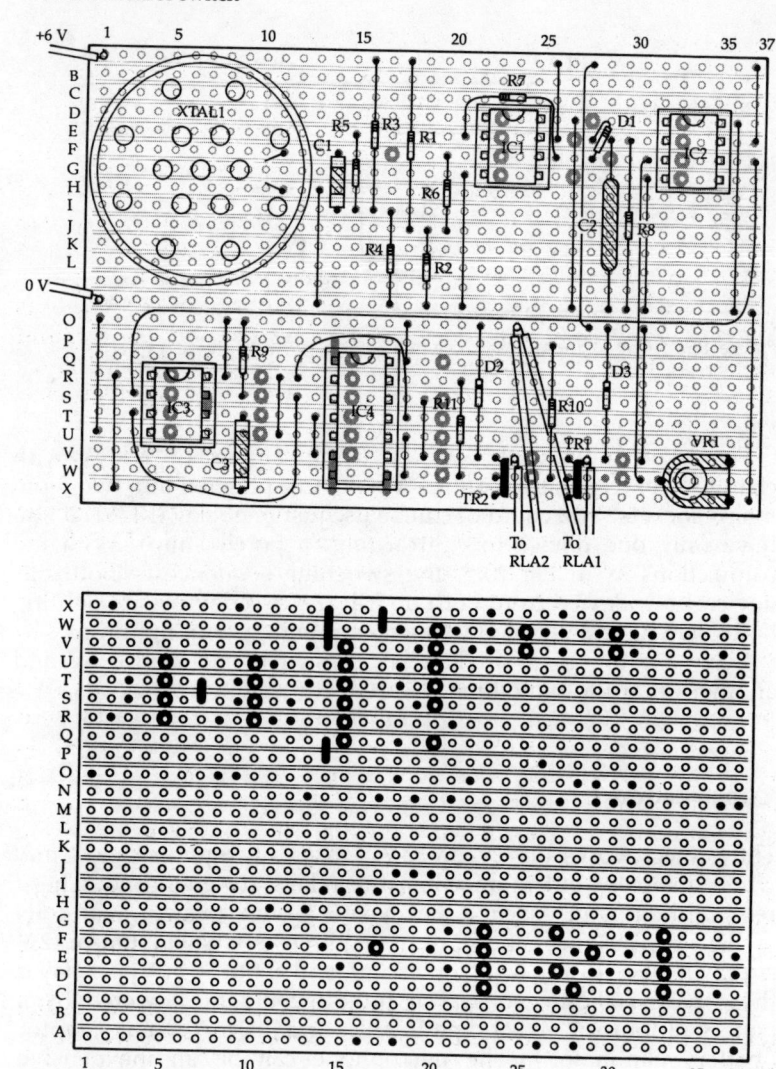

Figure 10.7 Stripboard layout of the remotely controlled switch

Build the circuit around IC1 first, including the microphone, D1 and C1. The voltage across C1 is between 2 V and 4 V, depending on the exact values of the resistors. Tapping the microphone causes the voltage to vary, but changes are rapid and you will

84

probably not detect them unless you have an oscilloscope. Continue with IC2 and its associated components. Adjust VR1 carefully - taking care not to vibrate the microphone or make any undue noise - until the output at pin 6 of IC2 *just* goes high (5.5 V or more). Now clap your hands, snap your fingers or whistle and the output falls to 0 V for an instant. The system is sensitive enough to show this effect at distances of several metres. VR1 is a sensitivity control. If VR1 is set so that the voltage at its wiper is about 0.25 V above that at pin 2 of IC2, the level of sensitivity is suitable for a typical room with an average amount of background noise. Setting the voltage only 0.1 V above pin 2 increases sensitivity at the cost of having the device triggered unintentionally by sundry clattering in the room.

When the monostable circuit (IC3) is added, its output is normally low but goes high when a clap is detected. The counter stage (IC4) is best tested using a voltmeter or the logic level probe (Project 1), or by temporarily wiring LEDs between the Q outputs of IC4 and the 0 V line. Repeated clapping runs the LEDs through the binary sequence from 00 to 11. If the counter jumps steps in the sequence, try increasing the value of C3 to lengthen the pulse from the monostable.

Finally wire up the transistor switches and relays. The relays are mounted on a separate board (Fig 10.8 and 10.9). The exact layout depends on the types of relay used. The boards are mounted on two bolts. Test the operation of the relays before making connections to the relay contacts. If the relays are to be used for switching low-voltage currents, there are few problems in wiring up. For general-purpose switching, each terminal of the relay is connected to a corresponding terminal or socket on the outside of the case. This allows the relays to be used for switching devices on or off when energised, and for the relays to be connected as reversing switches, if required.

For special purposes, you will use only certain terminals of the relays and wires from these could run direct to the devices being switched.

If the relays are used for mains switching, several points should be considered:

(1) The relays must be rated to withstand the mains voltage and current expected.

(2) Wires or other conductors carrying mains current must be kept well away from the other sections of the circuit.

(3) Take care that exposed metal parts carrying mains current can

Remote-control switch

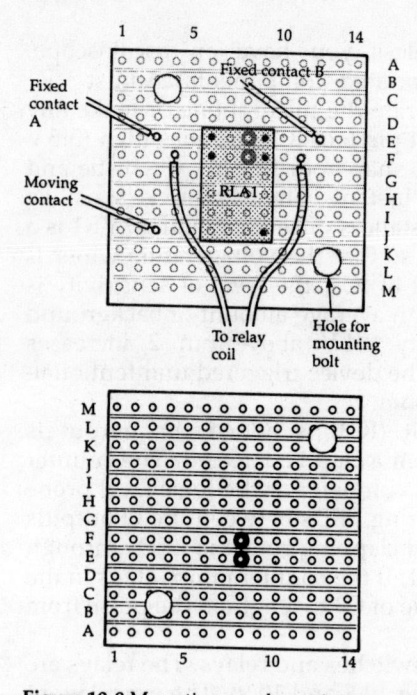

Figure 10.8 Mounting an ultra-miniature single changeover relay on stripboard; the connection to fixed contact A may not be required

not accidentally come into contact with other parts of the circuit when the project is finally inserted in its case.

(4) Bolt the relay board securely and, if the case is metal, use nylon nuts and bolts and include collars on the bolts to hold the board clear of the case.

(5) If the case is made of metal, the earth line must be securely connected to it. Solder a tag to the earth line and bolt this to a bare metal area of the case.

(6) Use insulated wire rated to carry the expected current; in general, do not rely on the copper strips of stripboard as conductors.

(7) When testing, take the mains supply from a socket fitted with a residual current device (RCD). This cuts off the current in fault conditions and is an essential precaution.

(8) Before connecting the device to the mains, check with an ohmmeter that there is no short-circuit between any of the 3 mains

Construction

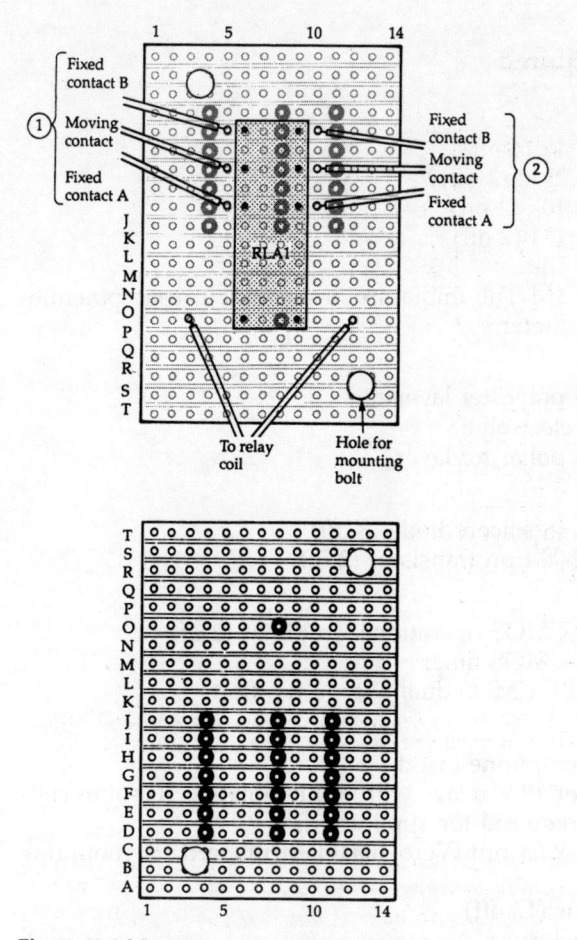

Figure 10.9 Mounting a minature double-pole mains relay on stripboard; connections to fixed contact A may not be required

lines inside the project, or between the live or neutral lines and the case.

(9) Close the case and fasten its lid *before* connecting the project to the mains. If a fault develops, switch off the mains and *unplug the project*, before attempting to open the case.

(10) If in any doubt, ask an experienced friend to help you.

(11) If still in doubt, try a different project.

Components required

Resistors (see p.8)
R1,R2, R10, R11	1k (4 off)
R3,R4	2M2 (2 off)
R5,R6	10k (2 off)
R7,R9	1M (2 off)
R8	18k
VR1	1M sub-miniature horizontal preset potentiometer

Capacitors
C1	100n polyester layer
C2	10 μ electrolytic
C3	220n polyester layer

Semiconductors
D1-D3	1N4148 silicon diode (3 off)
TR1,TR2	ZTX300 npn transistor (2 off)

Integrated circuits
IC1, IC2	7611 CMOS operational amplifier (2 off)
IC3	7555 CMOS timer
IC4	4013BE CMOS dual D-type flip-flop

Miscellaneous
XTAL1 crystal microphone insert
RLA1, RLA2 6V or 12V relays with approximately 200-ohm coils (contact types as required for application)
stripboard 95 mm x 63 mm (Vero 10346), stripboard for mounting relay(s)
1 mm terminal pins (13 off)
suitable plastic enclosure
8-way d.i.l. sockets (3 off)
14-way d.i.l. socket
bolts, nuts and washers for mounting relay board
plugs and sockets or terminals for making connections to controlled devices

11 Metronome

This is a device to appeal to the musician. It produces a regular beat at a rate that can be set from 40 beats to over 200 beats per minute. It accentuates every 2nd, 3rd, 4th, 5th, 6th or 8th beat. As well as its audible signal it has a pair of LEDs, a red one which flashes in time with the beat and a yellow one which flashes on the accentuated beat. The pitch of the audible signal can be adjusted according to the preference of the user.

How it works

The beat is derived from an astable multivibrator (IC1, Fig 11.1) running at a rate that can be set between 0.67Hz (40 beats per minute) and 3.47 Hz (208 beats per minute). This covers all the musical *tempi*, from adagio to presto. Actually the range obtainable is wider than this, since the design has to allow for tolerances in component values, particularly the capacitor. The multivibrator drives a pulse generator, that produces a high pulse at every beat. The length of this pulse is about 10ms, and appears at pin 1 of IC3. At each pulse, the red LED D1 flashes to mark the basic beat. The pulse passes on through two NAND gates, the purpose of which will be explained later, and arrives at pin 9 of IC5. Here it is NANDed with the output from the second multivibrator, IC6. This multivibrator produces an audio signal that can be adjusted to obtain a note of suitable pitch. Normally the input at pin 9 of IC5 is low. The signal from the multivibrator can not pass through the gate (see truth table, p. 3). When a pulse is present, the input at pin 9 is high. The (inverted) audio signal appears at pin 3 for the duration of the pulse. This is fed either to a crystal earphone or to a

Metronome

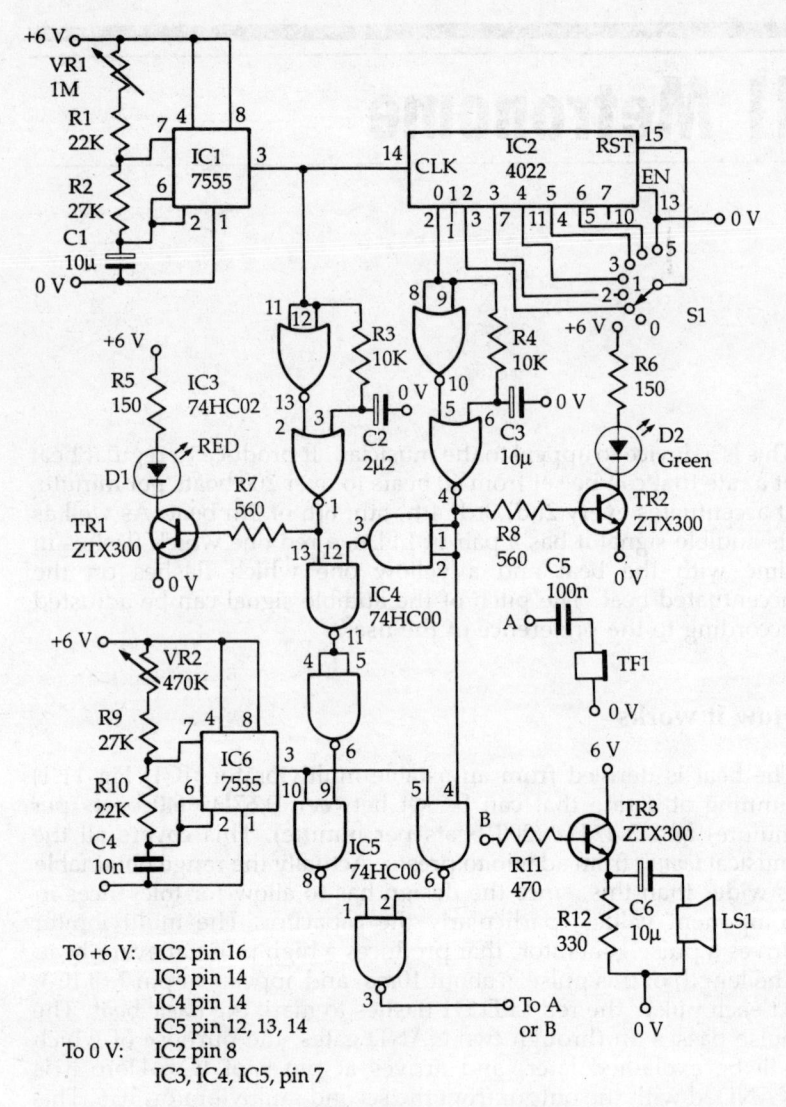

Figure 11.1 Circuit diagram of the metronome

1-transistor amplifier and loudspeaker. The result is a very short burst of sound, reminiscent of the 'tick' of a mechanical metronome. If you would prefer a 'beep' rather than a 'tick', the pulse must be lengthened by reducing the value of R3 to, say, 5k6 or 6k8 ohms.

The output from IC1 also goes to an 8-stage counter with 1-of-eight outputs. The action of this is described on p. 16. The rotary switch S1 allows the counter to be reset every 2, 3, 4, 5, or 6 counts or to cycle though 8 counts without resetting. Every time output 0 goes high the second pulse generator produces a pulse. Since C3 has a higher capacitance than C2, this pulse is longer (about 40ms) and is used to mark the accented beat. The accent pulse makes the green LED D6 flash. At the same time it gives a high input at pin 4 of IC5, allowing the audible signal from IC6 to pass through the gate. The result is a 'beep' lasting about 40ms, which sounds every 2nd, 3rd, 4th, 5th, 6th or 8th beat, depending on the setting of S1.

It is important that the basic 'tick' should not be heard on the accented beat. This is achieved by the gates of IC4. The output from pin 4 of the long pulse generator is fed to a NOT gate (a NAND gate with its inputs, pins 1 and 2, wired together). When the long pulse is absent, the output of this gate is high and the short pulses can pass through from pin 13 to pin 11 of IC4. In doing so, they become inverted, so there is another NOT gate (pins 4 and 5 wired together) to re-invert the signal. This ensures that the input at pin 9 of IC5 is normally low, preventing the audio signal from passing except when the short pulse is present, as already described. During a long pulse, pin 12 of IC4 is low and the short pulse can not pass through the gate. Only the long pulse reaches the earphone or loudspeaker.

Construction

Begin with the two multivibrators (Fig 11.2) Check the signals they produce by using a crystal earphone (of the type used with a transistor radio set) in series with a 100n capacitor, as at A in Fig 11.1. IC1 produces an audible 'double-click' at a rate which varies as VR1 is turned. IC6 produces a tone that varies in pitch from about 250Hz (about an octave below middle C) to about 2 kHz (about 2 octaves above middle C), as VR2 is turned. Now wire up IC2 and IC3, omitting S1 for the moment and temporarily wiring pin 15 of IC2 to the 0 V rail. Also connect the two LEDs. These can be mounted on the board, as in Fig 11.2, in which case you will need to cut an aperture in the lid of the case through which they may be viewed. Alternatively, mount them on the lid (as in Fig 11.3) and run wires to pins at U29/V29 and U31/V31. You need to include the wires from E27 to L28 and from B27 to M33 to make the LEDs function. The counter goes through its normal 8-stage cycle and the green LED flashes once for every 8 flashes of the red LED.

Metronome

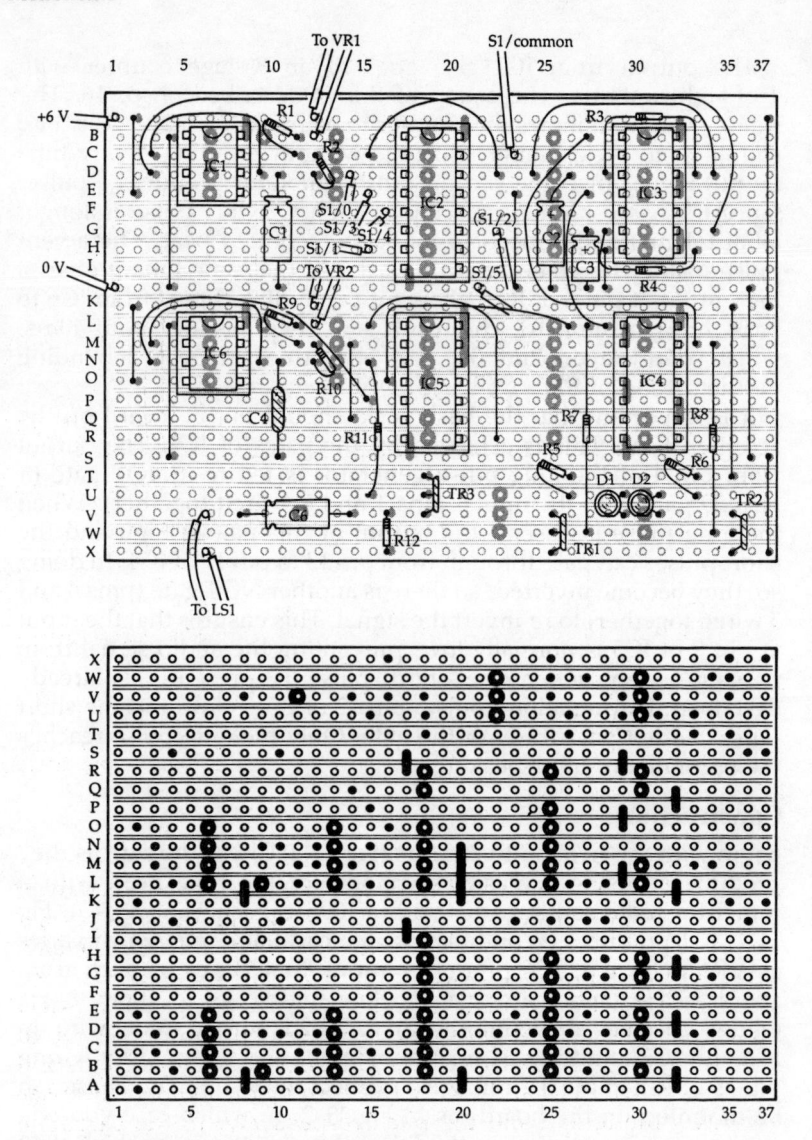

Figure 11.2 Stripboard layout of the metronome

Assemble the gating logic (IC4 and IC5). Note that several of the copper strips beneath these ICs are *not* cut, and there is no cut at Q25. The unused gate of IC4 is connected to pin 5; the unused gate of IC5 is connected to +6 V. Also connect either an earphone, as at

Construction

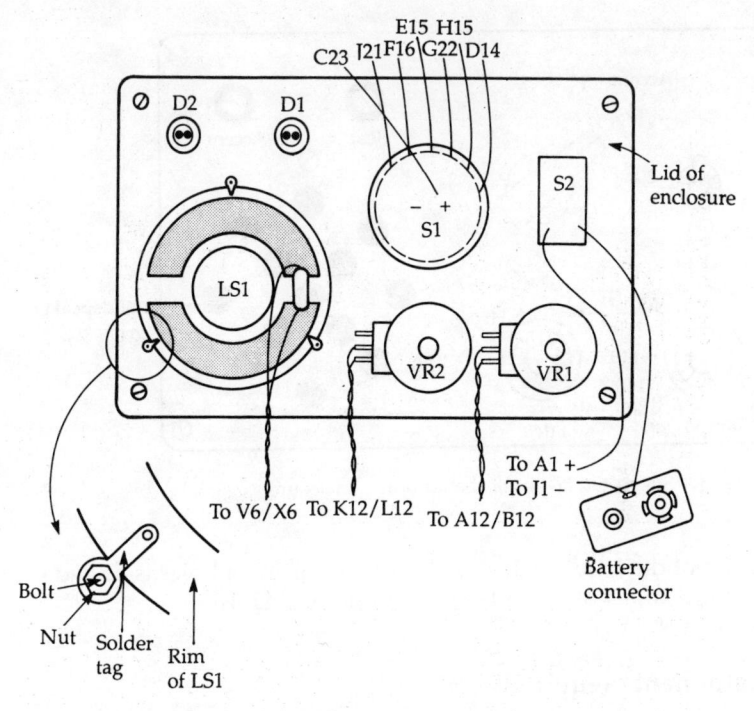

Figure 11.3 Suggested layout of off-board components, and off-board connections. Inset: securing the loudspeaker

A in Fig 11.1, or a loudspeaker, as at B. Fig 11.2 shows the layout for the loudpeaker version. The metronome produces a regular series of 'ticks' with a beep substituted for every 8th 'tick'. Finally, remove the temporary wire from IC2 pin 15 and wire the connections to S1. This is shown as a 6-way rotary switch. Most rotary switches are double-pole switches but only one set of contacts is used in this circuit. Fig 11.3 shows the connections and also how to wire VR1 and VR2 so that turning their knobs clockwise increases the tempo and the pitch respectively. In the earphone version, LS1 is replaced by a 3.5 mm jack socket to accept the jack plug on the lead of the earphone. The circuit requires an average current of less than 100 mA so it can easily be powered by a battery of 4 type AA cells, or possibly type C or D. It can be housed in a cheap plastic case, with the loudspeaker held in position by three bolts, each with a nut and solder tag (Fig 11.3). The solder tags grip

Metronome

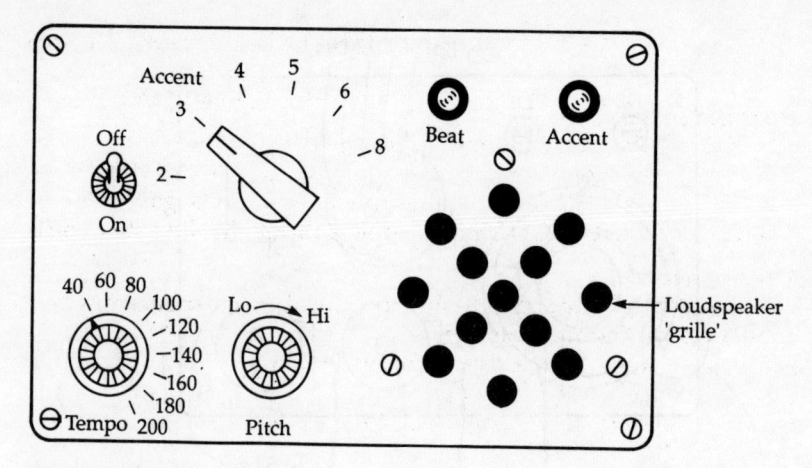

Figure 11.4 Suggested layout of the lid of the enclosure

the rim of the loudspeaker as shown. An array of holes is drilled in the panel to act as a loudspeaker grille (Fig 11.4).

Components required

Resistors (see p.8)
R1,R10	22k (2 off)
R2,R9	27k (2 off)
R3,R4	10k (2 off)
R5,R6	150 (2 off)
R7,R8	560 (2 off)
R11	470 loudspeaker version only
R12	330 loudspeaker version only
VR1	1 M carbon potentiometer
VR2	470 carbon potentiometer

Capacitors
C1,C3,C6	10 µ electrolytic (3 off, C6 in loudspeaker version only)
C2	2µ2 electrolytic
C4	10n polyester
C5	100n polyester - earphone version only

Semiconductors
D1,D2	TIL209 or similar LED (2 off, 1 red and 1 green)

Components required

TR1-TR3　　ZTX300 npn transistor (TR3 in loudspeaker version only)

Integrated circuits
IC1, IC6　　7555 CMOS timer (2 off)
IC2　　　　4022BE CMOS 8-stage timer with 1-of-8 outputs
IC3　　　　74HC02 CMOS quadruple 2-input NOR gate
IC4,IC5　　74HC00 CMOS quadruple 2-input NAND gate (2 off)

Miscellaneous
TF1 crystal earphone *or* LS1 8-ohm loudspeaker (inexpensive type, approximately 50 mm diameter)
S1　　　　double-pole 6-way rotary switch
S2　　　　SPST toggle switch
SKT1 3.5 mm jack socket (optional, for earphone)
knobs for VR1, VR2, S1
stripboard 95 mm x 63 mm (Vero 10346)
1 mm terminal pins (11 off)
suitable plastic enclosure
8-way d.i.l. sockets (2 off)
14-way d.i.l. sockets (3 off)
16-way d.i.l. socket
battery clip or 6 V battery holder
3 bolts, nuts and solder tags for mounting loudspeaker

12 Anemometer

An anemometer is a meteorological instrument for measuring wind speed. The usual form of anemometer has 3 cups mounted on a rotor (Fig 12.1). Wind makes the rotor turn at a rate proportional to the wind speed. The circuit of this project is one that measures the rate of rotation and hence the wind speed. The circuit is a simple *tachometer* and can readily be adapted for any application in which rates of rotation are to be measured.

Another form of anemometer has a propellor which rotates when the wind blows it. The rate of rotation of the propellor is proportional to wind speed so this circuit can also be used with this type of anemometer too. However, the construction of the mechanical parts of the propellor anemometer is more difficult. The anemometer shown in Fig 12.1 is not affected by the direction from which the wind blows, and operates just as well if the wind is rapidly changing in direction, but the propellor type has to be mounted so that it can swivel horizontally and has a vane to keep the propellor facing into the wind.

How it works

The rotor turns as the wind blows. The axle of the rotor has a disc on which is mounted a powerful bar magnet (Fig 12.1(a) and 12.1(b)). At each revolution of the rotor, the magnet passes close to a sensor ic. This sensor depends for its action on what is known as the Hall Effect. Inside the sensor there is a bar of semiconductor material through which a current is flowing. In the absence of a magnetic field, the charge carriers flow straight along the bar. If a magnetic field is present, the carriers are deflected to one side of the bar. This causes a potential difference to develop between one

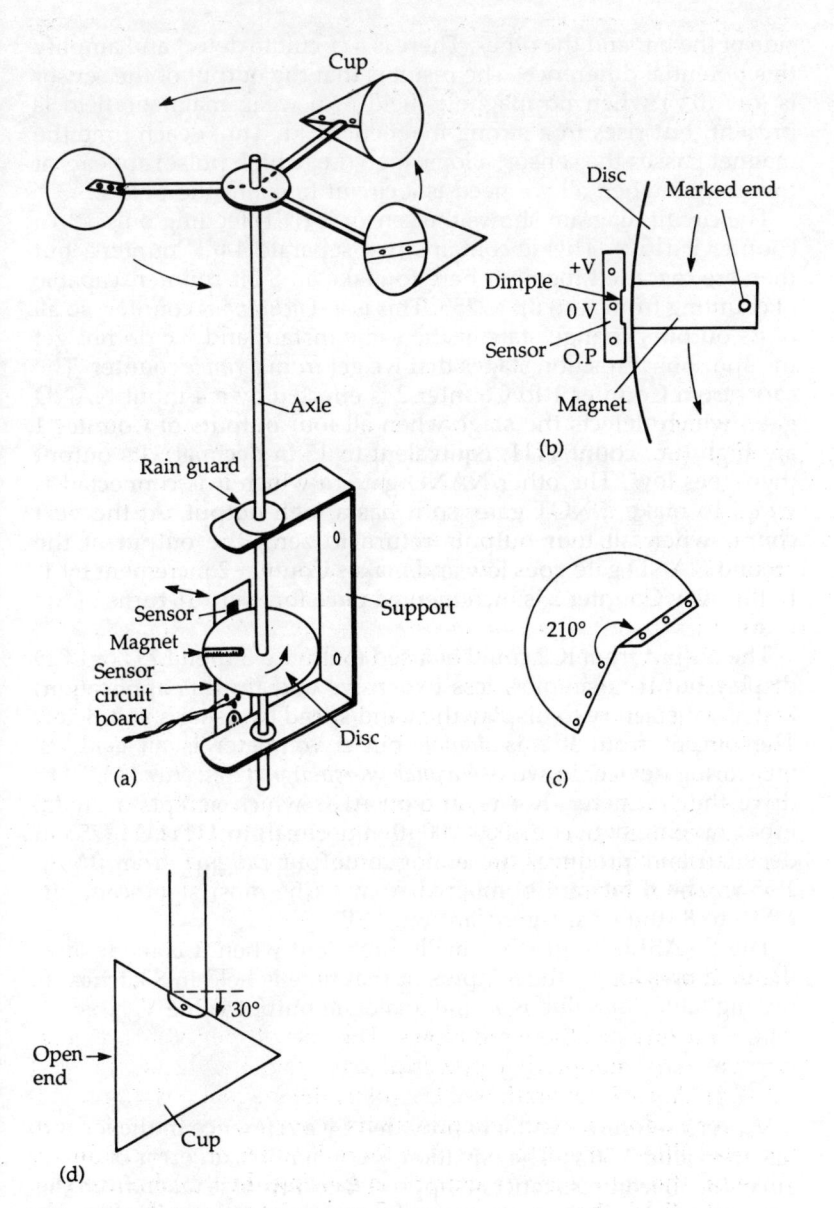

Figure 12.1 Building the anemometer

Anemometer

side of the bar and the other. There is a circuit to detect and amplify this potential difference. The result is that the output of the sensor is low (0 V) when no magnetic field or a weak magnetic field is present, but rises in a strong magnetic field. Thus, each time the magnet passes the sensor, a logic high (i.e. a high pulse) appears at the output. Then all we need is a circuit to count the pulses.

The circuit diagram shows the sensor ic (IC1) feeding pulses to a counter ic (IC2). This ic contains two separate 4-bit counters, but they are cascaded together here to make an 8-bit counter, capable of counting from zero up to 255. This is a *synchronous* counter, so all of its outputs change state at the same instant and we do not get the spurious transition stages that we get from a *ripple* counter. The carry from Counter 1 to Counter 2 is effected by a 4-input NAND gate, which detects the stage when all four outputs of Counter 1 are high (i.e. count 1111, equivalent to 15 in decimal). Its output then goes low. The other NAND gate to which it is connected is wired to make a NOT gate, so it has a high output. At the next count, when all four outputs return to zero, the output of the second NAND gate goes low and makes Counter 2 increment by 1. In this way Counter 2 is incremented once for every 16 turns of the rotor.

The output from IC2 could be used to drive a 3-digit LED or LCD display but it is simpler, less expensive and (in this application) just as satisfactory to display the wind speed by using a voltmeter. The output from IC2 is *digital*, but a voltmeter is an *analogue* measuring device, so we use a *analogue-to-digital converter* (ADC) to drive the voltmeter. IC4 is an 8-bit ADC which accepts a digital input ranging between 00000000 (0 in decimal) to 11111111 (255 in decimal) and produces an analogue output ranging from 0 V to 2.55 V. The 8 bits are numbered from 1 (the most significant bit, MSB) to 8 (the least significant bit, LSB).

The ENABLE input is normally high, but when it is made low, the data present on the 8 inputs at that time is held in 8 latches. If the digital input value is n, the analogue output value V_{OUT} is:

$$\frac{n}{256} \times V_{REF}$$

V_{REF} is a reference voltage, provided by a reference in the ic, and has the value 2.56 V. The equation above ignores an error of about 1mV due to lead resistance. If appreciable current is taken from the analogue output, the voltage available at output pin is reduced. This is because the output side of the ic behaves as a voltage source V_{OUT} in series with a 4k resistor. For accurate readings, we should

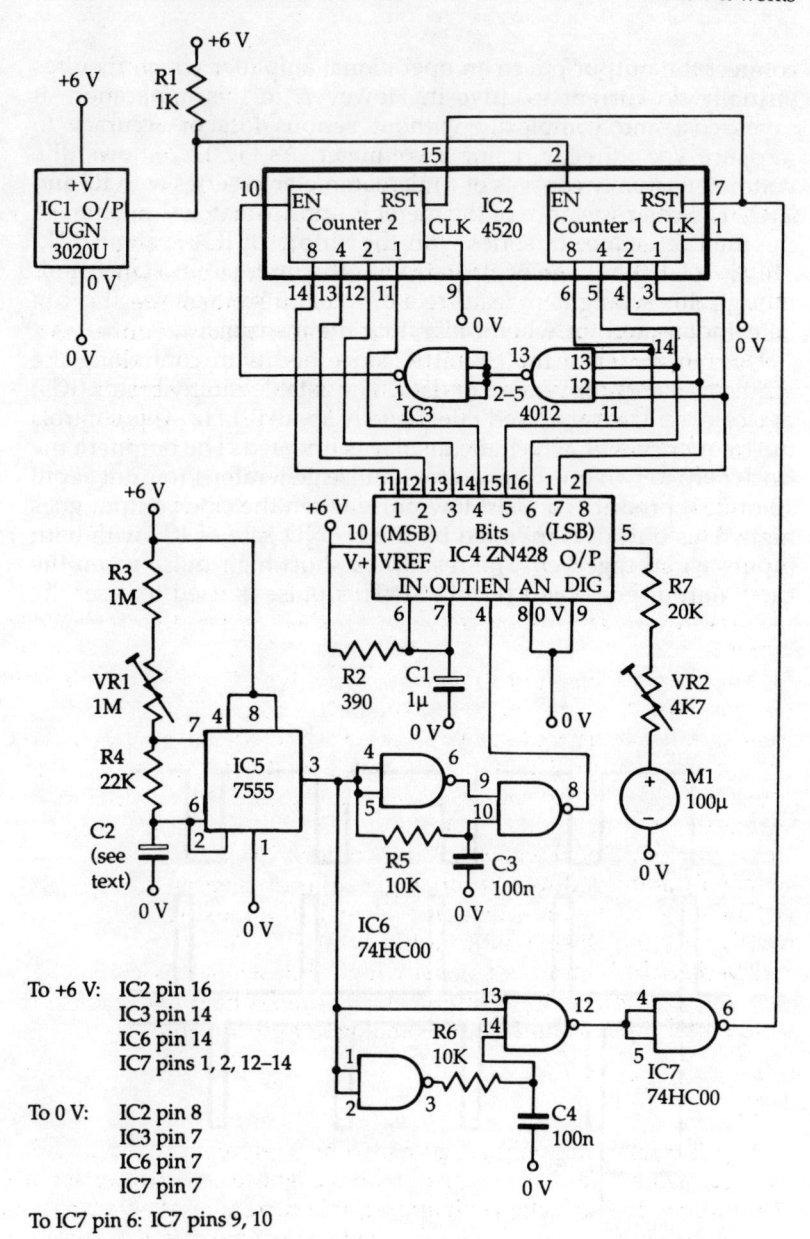

Figure 12.2 Circuit diagram of the anemometer

Anemometer

connect the output pin to an operational amplifier which requires virtually no current to drive it. However, in this application, it saves cost and complexity without serious loss of accuracy to measure V_{OUT} directly, using a voltmeter. As Fig 12.2 shows, the voltmeter actually consists of a microammeter in series with R7 and VR2. The coil resistance of the meter is approximately 1 kilohm, so the total resistance in series with the output of IC4 is about 23 k. This is relatively large compared with the internal resistance and, though this arrangment leads to a drop of output voltage, this can be compensated for when calibrating the instrument.

The remainder of the circuit is concerned with controlling the acquisition and conversion of data. The astable multivibrator (IC5) or clock oscillates at a fixed rate, usually about 0.1 Hz. This controls the frequency with which the display is updated. The output of the clock goes to two pulse generators. Pulse generator 1 (output at pin 12 of IC6) produces a short low pulse when the clock output goes high. This output is inverted by the NAND gate of IC7 with both inputs wired together. The result is a short high pulse when the clock output goes high (Fig 12.3). This pulse is used to reset the

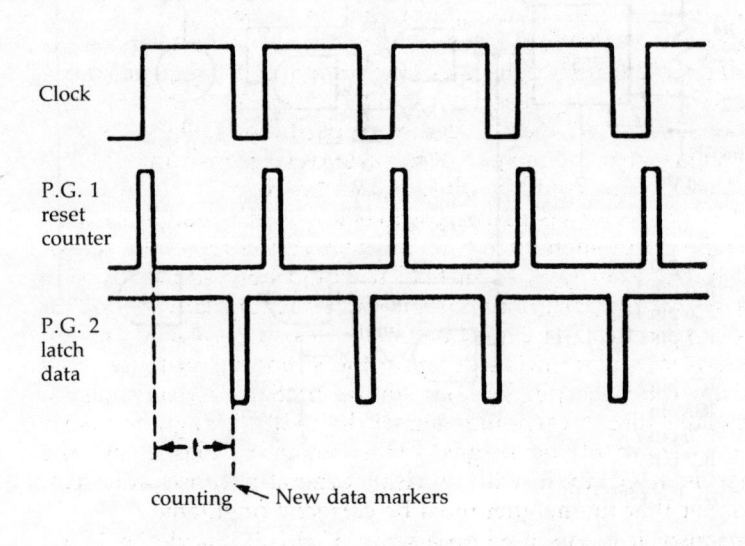

Figure 12.3 Timing diagram

counter. Counting begins as soon as the pulse is finished – virtually as the clock goes high. Pulse generator 2 (output at pin 8 of IC6) produces a short low pulse when the clock goes low. This is used to latch the data (the count) into the ADC. Thus the effective counting period t is almost equal to the length of time for which the clock output is high. Immediately the data is latched the voltmeter takes up the corresponding reading, which is held until the next time the clock output goes low. The scale of the voltmeter is marked in miles per hour or kilometres per hour and the wind speed can then be read.

Construction

The first stage is to build the rotor and its support (Fig 12.1). There are several ways of doing this, depending on the materials being used and the mechanical skills of the reader. In building the rotor it is necessary to compromise between light weight (so that it spins in the slightest wind) and robustness (so that it withstands the severest gale). The exact size is unimportant; cups can be about 5 cm in diameter, rotor arms about 5 cm long, and the axle about 20-30 cm long. The axle is best made from a length of steel rod, though wooden or plastic dowelling could be used. The cones are best made from sheet aluminium, cut as shown in Fig 12.1(c), curled round and bolted. The middle bolt is also used to attach the cone to the rotor arms. The end of the rotor arm is oblique as in Fig 12.1(d), so that the open end of the cone is parallel to the arm. The arms are fixed 120° apart to a disc of plastic or metal, which is secured at or near the top end of the axle. A large instrument knob from the junk box could possibly be used here.

Fig 12.1(a) shows a bent metal strip being used as a support. This part of the anemometer needs to be enclosed to protect the sensor circuit board from rain, so instead it would be possible to use a plastic box as support. Bore holes in the upper and lower ends for the axle to turn in. There must be small discs on the axle to prevent the axle from sliding up or down in the support, but these must allow the rotor to spin without undue friction. There is also a non-metallic disc to carry the magnet. Its radius is slightly greater than the length of the magnet. This can be of plastic and the magnet is fixed to it with a clear general-purpose adhesive. Remember that the magnet must be correctly orientated.

The sensor ic is mounted on a scrap of circuit board (Fig 12.4). The ic leads are left long to allow the body of the ic to be located in

Anemometer

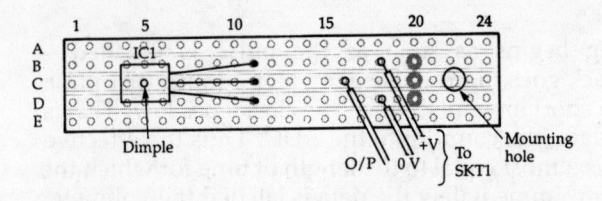

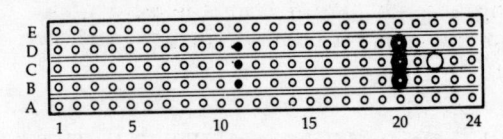

Figure 12.4 Layout of the sensor circiut board

the required position by bending them. Fig 12.1(b) shows the terminal connections, as seen *from above*. The gap between the end of the magnet and the dimpled face of the ic should be about 2 mm. Test the operation of the ic by connecting its 0 V and V+ terminals to 0 V and +6 V respectively. Measure the output voltage. This is normally very close to 0 V but rises almost to +6 V and then falls again to 0 V as the rotating disc brings the magnet past the ic.

Some thought must be given to weather-proofing. If the mechanism is enclosed in a plastic case, a disc mounted on the axle as shown in Fig 12.1(a) prevents rain from running down through the upper hole.

On the main circuit board begin with the timing circuits, the multivibrator and the pulse generators (IC5 to IC7, and associated components). For C2 use a 10 µF electrolytic capacitor. It may be necessary to substitute a different capacitor later. With VR1 set to its mid-way position the clock runs at *approximately* 0.1Hz. Use the pulse detector (Project 1) or an oscilloscope to check that pulse generator 1 produces a high pulse every 10s. Similarly check that pulse generator 2 produces a low pulse every 10s.

Mount IC2 and IC3, including the wires from D3 to J19 and from V27 to G23. When the rotor is turned by hand, the outputs of IC2 show an incrementing count, reduced to zero every 10 seconds. It is not easy to check the counting sequence fully but at least it is possible to show that the output at pin 3 changes at each turn of the rotor.

Complete the circuit by adding IC4 and the meter. In Fig 12.5, the holes for the 8 wires joining IC2 to IC4 are marked with the bit numbers. Join 1 to 1, 2 to 2, ... , 8 to 8. Try turning the rotor at various speeds. The meter reading changes every 10 seconds, increasing as the rate of turning increases, falling to zero if the rotor is not turned during the whole of a 10-second period.

The completed circuit can be housed in a plastic case and powered from a 6 V battery. The power switch (S1) and meter are mounted on the front of the case. Join the anemometer to the circuit by a twin lead long enough to run from the permanent outdoor location of the anemometer to the permanent indoor location of the main circuit.

Calibration

The meter is to be calibrated so that, using the short clock period, the scale reading in microamps corresponds to the wind speed in miles per hour. Calibration is done in 2 stages:

(1) Scale factor: Remove IC6 and IC7 from their sockets. Push a thin wire into the socket of pin 8, IC6 and connect its other end to 0 V. This makes the ADC continuously show the analogue of its digital input. Push a thin wire into the socket of pin 6, and connect its other end to +6 V. This enables the counters; to reset the counters, briefly connect this wire to 0 V and return it to +6 V again. Reset the counters in this way. Turn the rotor a few times, checking the output pins of IC2 to see that the counters are incrementing correctly. Continue turning the rotor until bit 1 is high and all others are low (equivalent to 128, almost exactly half the maximum reading). Adjust VR3 until the meter needle is half-way along the scale (50 μA, equivalent to 50mph). Replace IC6 and IC7.

(2) Timing period: If the instrument is to have the best possible resolution, the timing period should be adjusted so that the rotor makes 255 turns per period when the wind is at maximum speed (100 mph). Obviously there is no point in waiting for a hurricane before calibrating. Ideally, calibration is done using a wind tunnel, where wind speeds are accurately known. The more practicable technique makes use of natural wind. On a day when there is a steady wind blowing, mount the anemometer outdoors in its intended final position, as high as possible above ground and as far as possible from obstructing features such as roofs and trees. If you

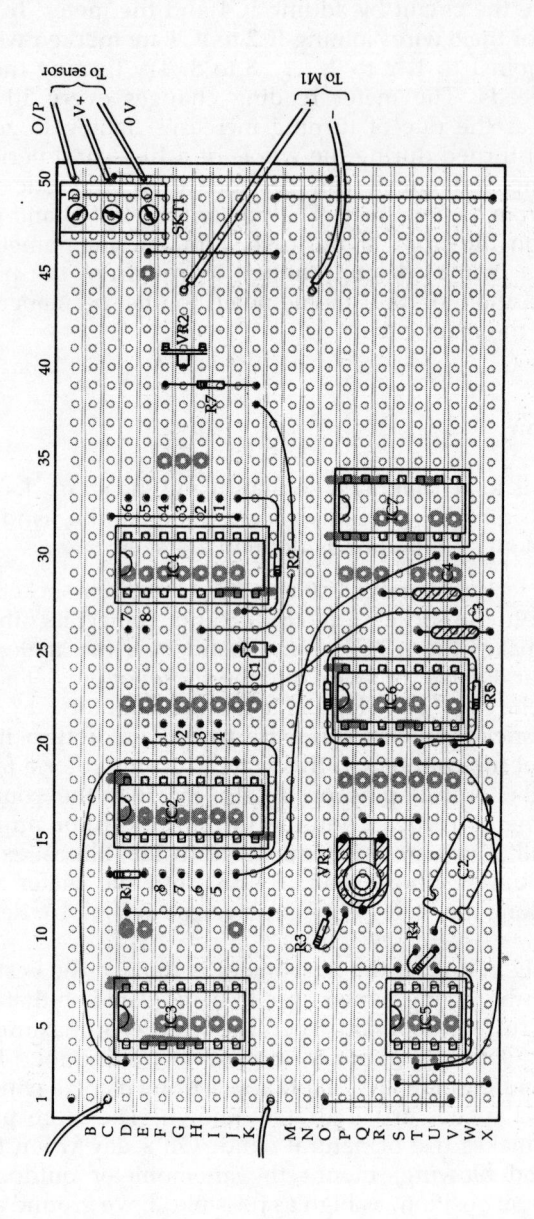

Anemometer

Calibration

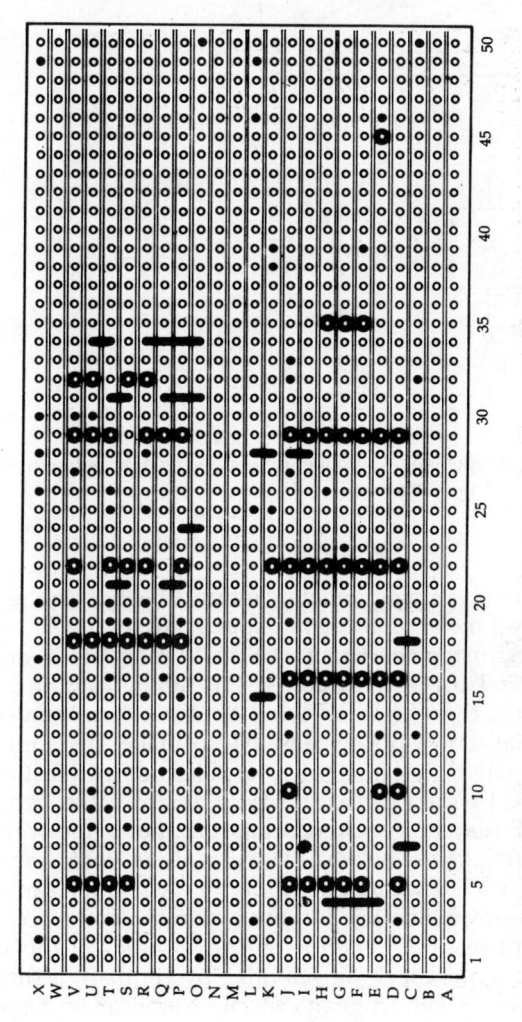

Figure 12.5 Stripboard layout of the anemometer (main board)

Anemometer

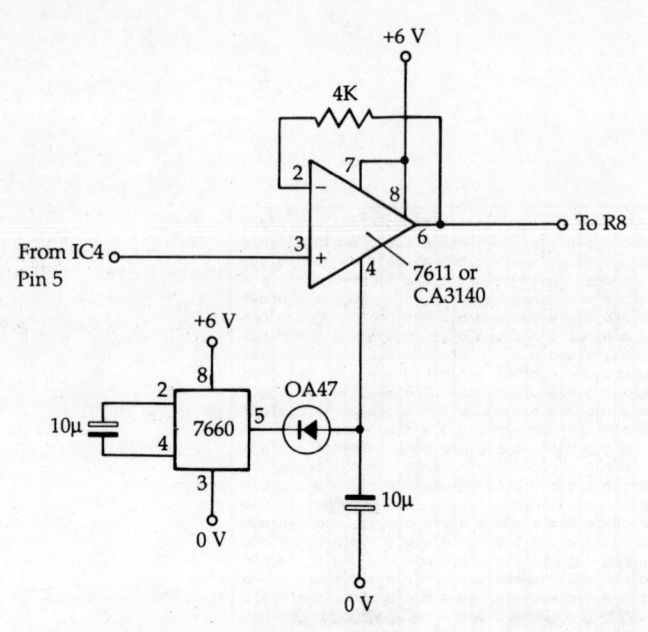

Figure 12.6 Using an operational amplifier to transfer the ADC output to the meter. Note the polarities of the capacitors

are able to borrow an anemometer, mount this beside the project anenometer and measure the wind speed. Alternatively, estimate the wind speed using the Beaufort Scale. Preferably it should be about 25 mph. Turn VR1 to adjust the timing period until the needle indicates the estimated wind speed when updated. It is quite likely that it will not be possible to make this adjustment. This will probably be because the rotor does not turn fast enough. If so, replace C2 with a capacitor of a different value. A value as high as 100 μF may be required, updating the meter every 100s. (but see *Modifications*, below). Having done this, repeat the adjustment of VR1. The meter reading in microamps can now be taken to represent wind speed in mph. However, the accuracy of any subsequent reading depends on the accuracy of this original estimate. You could check on readings by comparing them later with wind speeds taken from an official weather map made at the same time. Note that local wind speeds depend on land form and the presence of obstructions and may differ from those measured at the weather station.

Modifications

A way of increasing the rate of counting is to mount two or more magnets on the rotor. Although only one magnet is supplied with each ic, small magnets can easily be purchased. The type sold for use with reed relays is suitable. With, say, 5 of these mounted on the disc, the counter registers 5 counts for each rotation. This means that the updating time can be correspondingly reduced.

If you are using the circuit for other applications and wish for greater accuracy, an operational amplifier can be inserted between pin 5 of the ADC and R7. Fig 12.6 shows the circuit. There is room on the board for this. The ic requires a -6 V supply, which is obtained by using a voltage converter ic, as shown in the diagram.

Components required

Resistors (see p.8)
R1	1k
R2	390
R3	1M
R4	22k
R5,R6	10k (2 off)
R7	20k
VR1	miniature horizontal preset 1M
VR2	miniature vertical preset 4k7

Capacitors
C1	1 μ electrolytic
C2	see text (usually between 10 μ and 100 μ)
C3,C4	100n polyester (2 off)

Integrated circuits
IC1 UGN3020U Hall Effect switch and magnet (Electromail stock no. 307-446). The TL170C and TL172C ics are similar, but can *not* be used in this circuit as they have a latching action.
Additional magnets (optional, see *Modifications*)
IC2	4520BE CMOS dual 4-bit synchronous counter
IC3	4012BE CMOS dual 4-input NAND gate
IC4	ZN428 digital-to-analogue converter
IC5	7555 CMOS timer
IC6,IC7	74HC00 quadruple 2-input NAND gate (2 off)

Anemometer

Miscellaneous
M1 microammeter, 100 µA full scale deflection
S1 single-pole single-throw toggle switch
stripboard 127 mm x 63 mm (Vero 10345), scrap for mounting sensor
1 mm terminal pins (7 off)
suitable plastic enclosure
8-way d.i.l. socket
14-way d.i.l. sockets (3 off)
16-way d.i.l. sockets (2 off)
3-way terminal block, pcb mounting, 5 mm spacing
battery holder (6 V) or battery clip
materials for making the anemometer mechanism

Appendix: Notes for beginners

Stripboard

This is the circuit-building technique used in this book. Stripboard consists of an insulating perforated board with copper strips on one side. There are numerous variants in the arrangement of the strips, but we use the simplest type, which has continuous parallel strips running the length of the board. The strips are perforated with 1mm diameter holes, on a 2.5mm matrix. Boards are available in a range of sizes. For these projects we have nearly always specified boards in one of the standard sizes. However, it is often more economical to buy one of the largest size boards and cut it into smaller pieces. Terminal pins are available to fit the holes. The pins may be single-sided or double-sided; we use single-sided pins in these projects.

Using stripboard

If you are not using a ready-cut board of standard size, cut the board from a larger one, with a junior hack-saw. A fine flat file or glass-paper block is used to smooth the cut edges. In most projects some of the strips have to be cut, to separate parts of a strip that are used for different parts of a circuit. The layout diagrams indicate where the strips are to be cut. A special tool, called a *spot face cutter* is used for cutting away the copper strip around a hole. You can use a small (e.g. 3mm diameter) drill bit instead, but the proper tool is more convenient to handle. Usually a row of holes is cut to isolate one integrated circuit (ic) from another and to isolate pins on opposite sides of an ic. Occasionally, where a connection is required between one ic and another, or between

Appendix: Notes for beginners

pins on opposite sides of an ic, the strip is to be left uncut. Check the layout diagrams carefully to note such instances.

One of the commonest faults in circuit-building is to leave a thin hair of copper bridging a gap that is supposed to have been cut. Often the very edge of the strip remains uncut and bows out around the edge of the hole formed by the cutter. It is then very difficult to see, but it causes a short-circuit that has serious effects on the operation of the circuit. Every cut should be examined using a hand-lens. You may also find that flakes of copper remain on the cut ends of strips and may cause a short-circuit between the cut ends or to adjacent strips.

Nearly always the components are mounted on the plain side of the board, their leads being bent if necessary, passed through the holes and soldered to the copper strips (see later). It is preferable not to mount ics directly on to the board, since they are very difficult to remove once soldered in position. Also, it is useful to be able to remove them when a faulty circuit is being checked, or to avoid damaging them by static charges when adding further stages to a circuit. This is why we specify ic sockets for mounting ics.

Other connections on the board are made using insulated wire. For this purpose, single-stranded wire is by far the best; this is sometimes sold under the name of 'bell-wire'. A suitable gauge is 1/0.6 (i.e. one core, 0.6mm diameter). It is a good idea to buy a few lengths of wire of different colours, as the colouring often helps in sorting out the wiring of a complicated circuit. Multistranded wire (often sold as 'hook-up wire') is not suitable for use on stripboard, as stray strands tend not to pass through the holes, and may lead to short-circuits. Such wire is best for joining terminal pins on the board to off-board components.

There are several ways of mounting the completed circuit board, the method employed often depending on what type of case is being used. Some types of plastic case have slots moulded in the walls and the board just drops into these slots. Before mounting the components, check that the board fits neatly into its slots. If the case lacks slots, you can obtain pcb guides which consist of a plastic strip incorporating a slot, and with a self-adhesive base. These grip the edge of a board firmly and adhere securely to the wall of the case.

Another simple technique, suitable for small boards, is to use a lump of Blu-tack, or a double-sided adhesive pad, such as those sold under the name of 'Sticky Fixers'. Nuts and bolts may be used, and also self-tapping screws. Some cases have internal bosses specially intended for taking such screws. If you are using these, bore suitable holes in the board before commencing construction, taking care to locate the holes so that you do not cut copper strips that are being used as connections, and that the metal bolt and nut do not cause short-circuits between adjacent strips. It may be necessary to thread spacers or collars on the bolts to hold the board away from the case, especially if the case is made of metal. Alternatively, use nylon nuts and bolts, or mount the board on nylon 'stand-offs'. At one end a stand-off snaps into a hole drilled in the case and at the other it has a peg which fits snugly into a hole drilled into the board.

Soldering

Many people believe that soldering is difficult but, by employing the correct tools and following a few simple rules, it turns out to be surprisingly easy after a little practice. A mains-powered electric iron is best and you will need a mains socket close to your working area. For stripboard construction, it is essential to have a miniature low-power iron. A 15W or 18W iron is powerful enough and to use one of higher wattage is to risk overheating. The bit of the iron should be cylindrical (not a point), cut obliquely at the tip and not more than 2mm in diameter. A holder for the iron is useful, though some have a hook on the handle by which the iron can safely be suspended beside the workbench. Use multi-core solder-wire, 40% lead and 60% tin, in 22 standard wire gauge.

These are the steps to successfully soldering a joint:

1 Make sure the two surfaces to be joined are clean; wipe off any grease, and remove any corrosion with abrasive paper.

2 Switch on the iron and wait for it to heat up. Touch the end of it against the end of the solder-wire; if the solder melts *instantly*, the iron is hot enough. Attempting to solder when the iron is not hot enough can lead to trouble; the join takes longer to make and, during the extra time taken, heat can be conducted to sensitive components and damage them.

3 Wipe the bit on a damp sponge to remove crusty pieces of old resin.

4 Touch the iron against the solder-wire and spread a thin layer of solder over the tip of the bit. Add a *little* more solder so that there is *just* enough molten solder on the tip to flow on to the surfaces and make good thermal contact with them.

Appendix: Notes for beginners

5 Holding the iron in one hand, touch the iron against *both* surfaces that are to be joined. With the other hand insert the end of the solder-wire into the crevice between the two surfaces. The solder melts and flows evenly and *easily* over *both* surfaces. Run enough solder into the crevice to obtain a coating of solder on both surfaces (but not so much as to make a large 'blob').

6 When both surfaces are well wetted with molten solder, remove the iron and the solder-wire and allow the joint to cool. Do not move or disturb the joint until the solder has solidified.

7 Inspect the joint to check that the solder has flowed more-or-less evenly on to both surfaces. If the solder has formed drops or beads (like water on a greasy surface) there may be no electrical contact made. This is a 'dry joint'. It can happen if the iron is not hot enough, or it is not pressed closely to both surfaces, or the surfaces are dirty. If this happens, re-make the joint.

8 Check the area of board around the joint to make sure that you have not accidentally made hair-like bridges of solder between adjacent copper strips.

The essence of good soldering is a suitably hot iron and a quick easy action – step 5 above should take only 2–3 seconds.

Semiconductor components may become damaged if overheated during soldering. Use a heat shunt to prevent this. A heat shunt is a spring clip with jaws made of thick copper. It is clipped on to the wire leads of the component between the component itself and the part of the lead that is being soldered. Heat passing along the leads is shunted into the copper jaws, instead of going to the component. Remove the shunt immediately the joint has been soldered.

In several of the projects we deliberately use a blob of solder to connect two adjacent tracks. This is done to simplify the wiring. Check the stripboard layout drawings to ascertain where such blobs should be formed. Melt a blob on each of the two strips then hold the iron in both blobs, running in more solder to bridge the gap.

Some useful tools

In addition to a soldering iron, a junior hacksaw, a spot face cutter and a few assorted small screwdrivers, the following tools are particularly useful:

   (a) wire cutter and stripper: it is very difficult and exceedingly time-consuming to manage without a wire stripper; a cheap one is almost as good as the more expensive variety.

(b) a cutter for snipping off wires close to the circuit-board; these help make the board neat and get rid of short wire ends that may subsequently become bent over and make unwanted contacts.

(c) tweezers, for holding small components, inserting wires and a host of other tasks.

Which way round?

Fig A.1 shows the terminal connections of components used in the circuits in this book. See Figs 0.3–0.5 for ics.

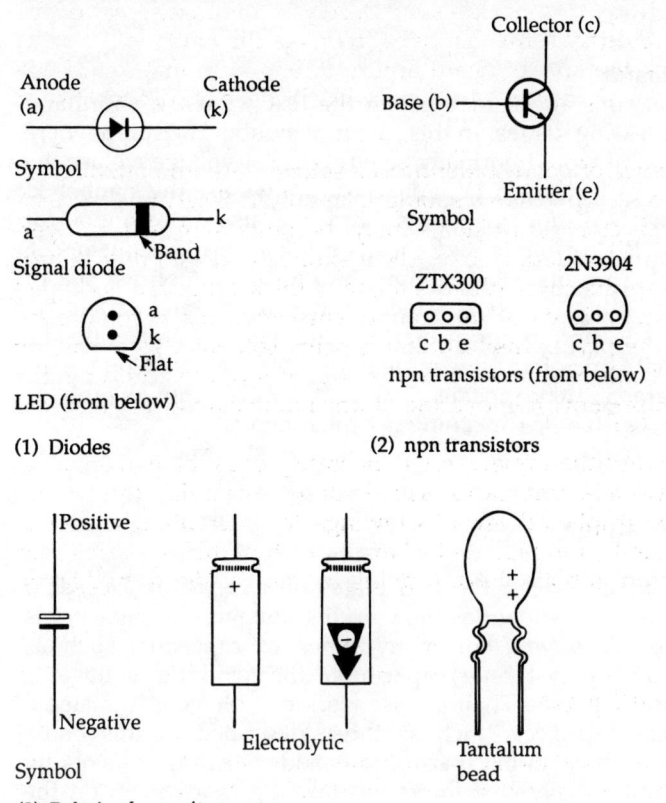

Figure A.1 Polarities

(1) Diodes

(2) npn transistors

(3) Polarized capacitors

Appendix: Notes for beginners

Units and values

The following electrical units are used in this book:

(a) Voltage or potential difference: The unit is the volt (V). 1V = 1000 millivolts (mV)

(b) Current: The unit is the ampere, more generally known as the amp (A). 1A = 1000 milliamps (mA). 1mA = 1000 microamps (μA).

(c) Resistance: the unit is the ohm (Ω). 1000ohms = 1 kilohm (kΩ). 1000kΩ = 1 megohm (MΩ).

(d) Capacitance: the unit is the farad (F), but this is such a large unit that it, or the millifarad, are hardly ever used. 1F = 1 000 000 microfarads (μF). 1μF = 1000 nanofarads (nF). 1nF = 1000 picofarads (pF).

Unit shorthand

On the diagrams and sometimes in the text we use a shorthand way of expressing values. In this, the unit symbol is written where a decimal point would normally be written. For voltage we use the symbol 'V'; 5V1 means 5.1V. For resistance we use the symbol 'k' for kilohms, 'M' for megohms, and we use no symbol (or sometimes 'R') for ohms: 5k6 means 5.6 kilohms, 56k means 56 kilohms, 5M6 means 5.6 megohms, 5R6 means 5.6 ohms, 56R or 56 means 56 ohms. For capacitance we use 'μ' for microfarads, 'n' for nanofarads, and 'p' for picofarads: 2μ2 means 2.2 microfarads, 22μ means 22 microfarads, 47n means 47 nanofarads, 820p means 820 picofarads. 100n means 100 nanofarads, though this is sometimes written 0μ1 meaning 0.1 microfarads.

Working voltage

If the potential beween the two plates of a capacitor exceeds a certain maximum, known as the working voltage, the capacitor is likely to break down. For many types of capacitor, such as polyester and polystyrene capacitors, the working voltage is several hundred volts. Their precise working voltage is immaterial in low-voltage projects such as those described in this book. Electrolytic and tantalum capacitors are made in a range of working voltages, and usually cost more and take up more room on the circuit board if they are rated for a high voltage. Electrolytic capacitors with working voltages of 10V, 16V or 25V are the best suited to these projects. The best are the 10V types, since these are usually the smallest and cheapest. Also it is preferable for an electrolytic capacitor to be operated at a voltage as close as possible below its working voltage. With tantalum bead capacitors the working voltage is usually 10V or 16V, with 25V or 35V for larger-capacitance types. If for any reason you are operating a circuit on, say, 12V instead of the 6V specified in the instructions, check that you do not use capacitors with 10V working voltage.

Power supplies

Most projects are best powered from a 6V battery. This can be a PP1 battery or even one of the PJ996 'lantern' batteries. Otherwise, the battery is made up from 4 dry cells in a plastic battery box or battery holder. Boxes are available for AA, AAA, C and D type cells, and are cheap.

The PP1 and many types of battery box have twin press-studs for connecting the battery to a 'battery clip', which has a complementary pair of press-studs. The battery clip has twin leads, usually red (+) and black (0V) for soldering to terminal pins on the circuit board. You may decide to instal a power switch in the positive lead. Usually a low-voltage toggle switch of the single-pole single-throw type is ideal. You could instead use a slide-switch or rocker switch. An advantage of the toggle switch is that it is mounted on the case simply by drilling a circular hole, whereas the other types usually require rectangular holes which are more difficult to finish tidily.

For projects that have more than minimal current requirements and that are to be operated for many hours or days at a time, it is better to use a mains-powered supply unit. It is possible, though not recommended, for the beginner to construct a PSU from individual components. It is possibly cheaper and certainly much more convenient to buy a ready-made PSU. For the projects in this book, which require no more than a few hundred milliamps at most, the ideal PSU is the mains adaptor, often called a 'battery eliminator', sold for powering portable transistor radio sets and tape recorders. These plug into a mains socket and can usually be switched to give a range of voltages, from 3V to 12V DC. For many projects the cheapest version with an unregulated output is suitable, but, if the current required is likely to approach the maximum that the unit can deliver (usually 300mA), then choose the more expensive model with a regulated output. Before purchasing, check that the output is DC (direct current), *not* AC (alternating current), since both versions look similar.

Cases

In describing the projects we have not specified the case or enclosure in great detail. This is because the type of case used is so often a matter of personal preference. The electronics buff may be content to leave a project uncased – just the bare board – and go on to experiment with a new project. Someone who has takes a pride in producing a unit with a really professional appearance may decide to buy a stylish two-tone case with anodised aluminium panel, tilt-legs and an integral battery holder. The better quality cases certainly have a fine appearance but are undeniably expensive and may be the major item of expenditure. In theory, any plastic sandwich box or any 'tin' of suitable size and shape may be pressed into service. But 'tins' tend to buckle when holes are bored in them, and many types of plastic crack or shatter when being worked. Most readers will probably strike the happy mean by choosing a case of basic pattern, made from a.b.s. plastic. Usually the off-board components are mounted on the lid of the case, or the panel, if there is one. An example is the metronome (Fig 11.4). Most off-board components are mounted in circular holes. Holes for LEDs and for passing wires through the case wall may be drilled with a bit of suitable diameter. Holes for most other components are usually larger. A set of round files is useful for enlarging drilled holes. If you are taking up the hobby seriously, a set of chassis punches is an investment that will save hours of time in the end. These are are the best for producing neat, round holes precisely where required. They will punch both sheet aluminium, tinplate and, taken slowly, a.b.s. and similar plastics. Chassis punches are also available singly; the 9.5mm (3/8") diameter punch is one that is likely to be used most often.

Magazines and catalogues

It is strongly recommended that you take at least one magazine regularly and buy one or two component catalogues from the larger mail-order electronics houses. The catalogues are usually mines of information about components and contain masses of data and often useful circuits too. Together, magazines and catalogues represent the best way of keeping in touch with your hobby and learning more about it.

Index

analogue-to-digital converter, 98–100
AND operation, 2–3, 60
astable multivibrator, 16–17, 25–26, 30, 61
audible warning device, 27, 36

binary system, 8
bistable circuit, 11–12
Boolean algebra, 60

cases 116
CMOS, 7
comparator, 34
contact bounce, 51
counter, 16, 26, 98–99

data selector, 43
de-bouncing, 51
decimal system, 8
digital logic, 1
divider, 16

enclosures, see cases
EXCLUSIVE-OR operation, 3

false, see untrue
flip-flop, 11–12, 79–80
forward voltage drop, 9
frequency divider, 26

gate, 2

Hall effect, 96–98
handling ics, 7
heat shunt, 112

high logic level, 4

ics, see integrated circuits
integrated circuits, 4–6, 7
INVERT, see NOT

logic, 1, 7–8, 60–64
logic gate, 2
logic ics, 4–6
low logic level, 4

monostable multivibrator, 25, 27, 31
music ic, 68

NAND operation, 3, 62
NOT operation, 3, 11
numbers, 7–8

OR operation, 3, 62

phase-locked loop, 34
phototransistor, 72
PLL, see phase-locked loop
polarity, 113
power supplies, 115–116
pulse-generator, 42–43, 73–74, 99

races, 64
relays, 81–83, 85–86
resistor, 8
reversing relay, 82

Schmitt trigger, 51, 72
shift register, 67
soldering, 111–112

Index

stripboard, 109–111

tachometer, 96
timer ic, 24–27, 30–31
tools, 112–113, 116
true, 1, 16
truth table, 3

units, electrical, 114
untrue, 1

voltage levels, 4–5, 9, 114

wiring, 110
working voltage, 114